L. H Russwurm

Canada

A Geographical Perspective

Author

Louis-Edmond Hamelin
Laval University, Quebec

Translators

Margaret C. Storrie
Queen Mary College
University of London, England

C. Ian Jackson
Ministry of State for Urban Affairs
Ottawa

Wiley Publishers of Canada Limited
Toronto

Canada

A Geographical Perspective

The present text has been revised and enlarged by the author to include information not available at the time of original publication.

Published outside Canada by John Wiley and Sons (New York, London, Sydney).

Library of Congress Catalog Card
No. 72-12415

ISBN 0-471-34680-2

Photographs by:
Information Canada Photothèque 12, 17, 21, 34, 42, 45, 50, 60, 73, 76, 103, 106, 156, 160, 175
Dept. of Energy, Mines and Resources 61
Iron Ore Company of Canada 146
George Allen Aerial Photographs 148
Dept. of Travel Industry, Gov't. of British Columbia 185
Ministry of Transportation and Communications, Ontario 203

Designed by Christine Purden Associates, Toronto

Printed and bound in Canada by The Bryant Press Limited for Wiley Publishers of Canada Limited

Contents

Maps and Graphs

Plates

Tables

Preface

The distinctiveness of this book is due, in the first place, to the fact that it is the work of a single author. It therefore conveys a consistent view of geography, something that cannot be identified in works consisting of chapters that are methodologically quite distinct from one another. The reader, whether or not he is a geographer, may find this unified approach attractive.

Secondly, the book appears to be the first real geography of Canada produced in the country itself from a viewpoint that is other than English Canadian. It is, however, neither a geography of Quebec nor a survey of Canada from the perspective of one of the ten provinces. At a time when questions of national identity continue to be of concern, such a book, intended as an objective study, may be of some interest. It may also encourage the synthesis of our two "national" views of geography; in such a case its interest may extend to historians of geography and of scientific literature.

No rash claim is made, however, that this text represents a comprehensive or complete geography of Canada. Five themes, and only five themes, are used in its construction: the influence of cold conditions, a geography of space, the different cultural groups, economic structure and, lastly, urban life. I have tried to draw general geographical conclusions by showing how these primary features of the country interact with one another. This leads naturally to a concluding section on the mental image of Canada. The work is therefore an exercise in systematic geography. An attempt is made to analyze such general characteristics of Canada as its vast area, its cold and its small population; this has entailed the better definition of notions of temperature range, northern character, ecumene, regional differences, and Canadianism.

Two points should be noted about the spatial dimension. Earlier writers tended to be preoccupied with a so-called useful Canada, located close to the U.S.A. border. Benefiting from recent research, I am concerned with the whole of Canada, including the northern part. As well as being a subject for geographical study in its own right, the North exerts a great influence on southern Canada. It is in the North that the southern climate is in part determined. Northern areas are a source of minerals for the industries of "Base Canada". Politically the North is a reminder that the U.S.S.R. is becoming more and more a neighbour. Some coastal areas in the North are places where the adaptation of Eskimos to their environment is very highly developed. This book is not concerned, therefore, merely with the main areas of human settlement in Canada. Secondly, because of the distribution of land and water, the area relevant to a geography of Canada extends beyond the land areas within Canadian sovereignty; to the official area of the country must be added the inland and coastal water areas.

The present edition is more than a straightforward translation of the volume originally published in Paris in 1969 by Presses Universitaires de France. The core of the translation does reflect the aims of the editors of the series in which it first appeared. However, although much of the text has been deliberately retained from the original, very many changes, additions and updatings have been made. The result is therefore more a revised edition than a translation. In particular, 1971 census data have been incorporated where available, the bibliography has been completely revised and the illustrations are new and more numerous. Several diagrams have been added, including an isodemographic map. Finally, this edition contains a detailed index. The translators, the publishers and I have worked in close collaboration so as to produce a translation that we hope is both accurate and elegant.

I am very grateful to all those who have made the publication of this edition possible, especially the translators and those who provided the illustrations.

Louis-Edmond Hamelin
Quebec
August 1972

Translators' Note

Professor Hamelin's preface to this edition arrived for translation shortly before the manuscript was typeset. We then discovered that, to a very great extent, the reasons why we undertook the translation of the book were also those that led the author to write it. No further explanation seems necessary on our part and it remains for us to underline the importance that he attaches to the collaboration among author, publishers and ourselves. We have also a particular debt of thanks to Miss Jean Hepburn and Mrs. Joan Butler for their secretarial help. Which was the more difficult task, transcribing the first rough draft from tapes, or deciphering the thousands of pencilled amendments to that draft, we dare not inquire.

Margaret C. Storrie
C. Ian Jackson
London and Ottawa
October 1972

Introduction

Immense size is one of the most striking geographical characteristics of Canada; its area is exceeded only by its northern neighbour, the Soviet Union. Canada is about the same size as China or Europe and is larger than the United States or Brazil. Moreover, the spatial extent of Canada is even greater than is usually recognized. Two measures of its area are in fact possible. The first is that of the extent of the land and the inland waters, which totals 3,850,000 square miles (9,856,000 square kilometres). This is the official area of Canada, according to the statistical yearbook. There is also the geographical extent of "Greater Canada", adding the enormous central area of Foxe Basin, Hudson Bay and James Bay, the arctic channels and the Gulf of St. Lawrence. In addition there are the straits along the Pacific coast, and other coastal waters, including the Grand Banks off Newfoundland. This perspective is mentioned not as a form of political imperialism but because Canada's marine areas significantly influence her geography through climatic and glaciological factors, and also have considerable economic significance. Defined in this way, Greater Canada has an area of 4,800,000 square miles (12,248,000 square kilometres). Of this total, water represents 27 per cent, a more realistic proportion than the official figure of 7 per cent, which includes only inland waters.

Greater Canada is a territory which stretches for 3,000 miles (4,828 kilometres) from Lake Erie to the Arctic Ocean and from the Grand Banks off Newfoundland to the mountains on the boundary between Canada and Alaska. It has a land border of more than 5,500 miles (8,800 kilometres) with the United States. A distance of 6,500 miles (10,460 kilometres) separates the two ports of Vancouver and Halifax via Panama, and it is still farther via the difficult Northwest Passage. When it is 5:30 P.M. at Goose Bay on the Atlantic, it is only noon at Dawson in the Yukon. These huge dimensions are a measure of the greatest problem of the country, sheer distance. This expresses itself as an economic disadvantage, and also as a factor which has influenced the history of settlement. The enormous distances have discouraged close relationships between different parts of the country, and the cultural survival of French Canada must owe something to this factor. The diverse regional characteristics have led to the difficult problem of regional disparities. The elementary fact of the size of the country cannot be forgotten in the analysis of any Canadian problem. "There are few major national or regional issues in Canada which do not have significant geographical aspects."[1]

The location of Canada is such that, in physical terms, it has been characterized by both isolation and accessibility. It was only in the seventeenth century that the development of navigation enabled Europe to colonize permanently its Far West. Prairie wheat is at least 3,500 miles (5,632 kilometres) from Liverpool, and that for only three months of the year via Churchill. The Pacific coast of Canada is 7,000 miles from Singapore and a similar distance from Australia; Alaska separates the Yukon from the nearest part of the U.S.S.R. In all international trade, such vast distances, added to internal distance, work against Canada. But such isolation has also been a protection, and Canada has been able to take part in two world wars without its own territory being attacked, something which has been a great aid in its economic development.

It is not always a matter of simple distance: in the case of the U.S.S.R., for example, it is not

1. C. I. Jackson, "Wide Open Spaces on the Map of Canada", *The Geographical Magazine* XLIV (1972): 342–351.

so much the 1,200 miles (1,931 kilometres) between it and Canada that forms the real barrier as the polar climate, the language differences and the political and philosophical barriers. Nevertheless, distance from other countries has inevitably pushed Canada towards its powerful neighbour. Too far from the rest of the world, Canada has become very close to the United States.

In a latitudinal comparison, the position of southern Canada is surprising. Lake Erie is at the same latitude as Rome; the southern boundary of western Canada is on the line of latitude which passes through Paris. These southern latitudes enable southern Canada to have a good share of solar energy. But, on the other hand, the 60th parallel of latitude is largely uninhabited in Canada, but Oslo, Stockholm and Leningrad lie along it in Europe. In terms of population distribution, these northern latitudinal contrasts are due not merely to differences in the period of settlement; they represent fundamental climatic differences on either side of the Atlantic Ocean. Northeast Canada receives more imported cold from polar regions than do the Scandinavian countries.

This is the land that this book is concerned with.

Part One

Temperature Range and Northern Character

Beyond argument, the North is one of the basic characteristics of Canada. Much more than the Norden group (Norway, Sweden, Denmark, Finland and Iceland), Canada can truly be termed a northern country, since Ellesmere Island stretches almost 1,200 miles (1,900 kilometres) beyond the Arctic Circle and tundra covers more than a third of Canada's surface area. Through advection of polar air masses or movements of sea ice, even southern Canada is affected for part of the year by conditions which are unusually severe for the latitude. The Mistassini Indians have a saying, "Winter is the North." It is this environment which led the historian D. G. Creighton to coin the phrase "Dominion of the North", for a very large part of Canada, if not the whole of it, is either in the North or "of the North".[1] The north-pointing triangular leaf used as the symbol of the centennial of Confederation in 1967 expressed the northern character of Canada.

Chapter 1

Air Temperature

More than anything else, the North represents cold. Lack of heat is one of the basic elements that shape the geographical personality of Canada. In the south of the country, this is usually true of only one season; in the Far North, however, the cold persists virtually all year round and significantly affects three major features of the environment: the air, the ground, and the water areas. To a very large extent, therefore, forms of life are a function of heat potential.

It would be difficult to exaggerate the importance of the seasonal variation of temperature. Canada is a country of two seasons distinguished from each other not by differences in amount of precipitation, as they are in some tropical areas, but by temperature. Canada in summer is a very different place from Canada in winter; there are, in fact, two Canadas, defined by these main seasons. The relative durations of the two periods differ from south to north; in general, the duration of the "good" season shortens as latitude increases.

Indices

Canadian temperature characteristics can be appreciated through certain useful indices. Freezing or negative values represent the annual total of the differences between 32°F and the daily temperature of those days when the air temperature is below freezing. Conversely, thawing or positive values represent mainly summer conditions, the sum of the day-degrees

1. W. C. Wonders, ed., *Canada's Changing North*, p. 351.

above 32°F.[2] It goes without saying that negative values increase and positive values diminish the farther north one goes. There is a transition zone between a northern area where the annual balance is negative and a southern zone where the annual balance is positive.

These calculations indicate that Canada is not equally divided by the threshold of 32°F; the tendency towards winter conditions explains why similar isotherms—such as 3,000 day-degrees—are found farther to the south in the cold season than they are in the warm season. By advection, most of Canada gets more cold in winter than warmth in summer, whence, perhaps, the remark of the poet Gilles Vigneault: "My country is not a land; it is a winter."

The thermal balance along a meridian from Ellesmere Island to Lake Erie is only one of the indicators which might be used, and it does not do justice to the important regional variations in Canada east of the Rockies. In winter the isotherms are concave towards the Arctic Ocean, with Manitoba and Ontario near the 56th parallel being the provinces where the polar air reaches farthest south. In summer, by comparison with the Quebec-Labrador peninsula, a major shift in the latitudinal alignment of the isotherms benefits the Mackenzie Valley.

One result of this natural deficit of warmth is the need for heat in factories and buildings. In calculating these needs, a daily threshold temperature is fixed at 65°F (18.3°C). As before, the differences between this temperature and each day's temperature are calculated and summed. All parts of Canada require heating, even the Pacific coast. The deficit in Churchill is twice that in Montreal; that at Resolute, in the Far North, is four times that in Vancouver. The Ontario peninsula, which has a total deficit 1,000 day-degrees less than the Gulf of St. Lawrence and 3,000 less than the southern Prairies, has, therefore, a relative natural advantage. When examined month by month, taking Montreal as an example, the burden of winter is shifted towards spring: February has the same deficits as December; those in March are greater than those in November, in April greater than October, and in May greater than September. In Montreal, therefore, the heating season lasts for eight months and is intense for five.

Among the factors influencing temperature, there are several which favour the development or increase of negative values. The obliquity of the earth's axis means that for several months the polar regions are out of the field of direct solar radiation, and even during the other months the angle of incidence is low and receipts of solar energy are therefore limited.

Radiation

Energy is derived from the sun. According to various authors, about 10 per cent of solar radiation is directly absorbed in the upper atmosphere, 45 per cent is reflected by clouds and 45 per cent continues towards the surface of the ground. What happens to this last part depends on the type of surface which receives the radiation: the surface layers of air, water vapour, ice sheets, snow, sand, dark rock or plant cover may either absorb or reflect the solar rays. The *albedo*, high (about 88 per cent) in the case of a winter snow cover which is white and "dry", is a measure of the proportion which is reflected. These high energy-losses associated with a snow cover delay the onset of spring; when the snow disappears net radiation increases rapidly. Albedo, like *advection*, prolongs the winter. The earth, which thus absorbs only a proportion of insolation, is itself the source of longer wave radiation; of 100 Kcal per square centimetre received in Hudson Strait, only 20 are actually absorbed. At Kapuskasing,

2. H. A. Thompson, "Air Temperatures in Northern Canada with Emphasis on Freezing and Thawing Indexes", *Proceedings, Permafrost International Conference* (Washington, 1965), pp. 272–280.

net radiation is only 15 per cent of that which is received and is negative in the months of November, December and January. The loss of terrestrial heat energy, most apparent at night, serves to warm the lowest layers of the atmosphere and contributes to the coldness of valley bottoms. The energy balance between sun, earth and atmosphere is maintained through a complicated set of processes. The Canadian situation is characterized by the presence of a polar "vortex" which, despite the upper-level advection of tropical air, is always marked by lack of warmth.[3] This is the most important source of Canadian air masses.

Air Masses

The atmosphere may be conceived of as divided into sections or air masses which, in the horizontal plane, are characterized by approximately uniform temperatures and humidities. Each air mass, and each front separating two masses, gives rise to distinctive weather conditions. Surface weather observations do not always indicate the true nature of the air masses aloft, which must be determined by rawinsonde ascents. From north to south, four general air mass types affect Canada.

(1) *Continental arctic air* (*Ac*) is the coldest air mass; formed over the ice, snow and frozen ground of the Far North, it is also dry. Although in summer it is restricted to a part of the Arctic Archipelago, in winter it covers the major part of the Canadian land mass. It is seldom found on the west coast but it regularly affects the St. Lawrence Lowlands. The low total precipitation and modest snowfalls in the North are due to this air mass.

The three other air masses are classed as *maritime* because, by contrast with *Ac*, they contain moisture of oceanic origin.

3. F. K. Hare, "Recent Climatological Research in Labrador Ungava", *Cahiers de Géographie de Québec* 19 (1966): 5–12.

(2) *Arctic maritime air* (*Am*) has a moderate moisture content, leading to the development of various types of cloud. In winter it is developed from *Ac* air which passes rapidly over the North Pacific before reaching the continent. Similar characteristics can develop along the Atlantic coast, but such arctic Atlantic air does not penetrate far into the continental interior. In summer, humid air of arctic origin can develop even on a wholly continental trajectory, due to the major sources of moisture and evaporation (arctic straits and channels, Foxe Basin, Hudson Bay, James Bay, and innumerable lakes).

(3) *Polar maritime air* (*Pm*) is found farther to the south and is moister than arctic air masses; it represents a mass of air which has remained for a longer period at lower latitudes over neighbouring oceans. Since the principal source of such air is the Pacific, the general circulation leads to a strong contrast in precipitation totals between the coastal mountains and the Prairies. Because the presence of the mountains encourages the development of local precipitation, there is a major contrast between conditions on the western side of Canada and those in western Europe: the Prairies are practically cut off from precipitation of Pacific origin.

The general circulation of the atmosphere leads to an exchange, in a sloping vertical plane, between warm, moist air masses predominantly of tropical maritime origin and cold, dry air originating in northern continental areas. The convergence of the two takes place in mid-latitudes, especially in southern Canada. The three masses just described are found to the north of this narrow boundary zone known as the polar front.

(4) *Tropical maritime air* (*Tm*) develops in the Gulf of Mexico, the Caribbean and in the neighbouring oceanic areas south of 30°N. This mass is very warm and moist. In the lowest layers, *Tm* rarely penetrates beyond the St.

Lawrence Lowlands during winter. Its convergence with northerly airstreams leads to storms and snowfall. Its passing over cold surfaces such as the North Atlantic or even the Great Lakes leads to fogs which are a hazard to navigation. During summer, showers and cloudbursts occur at the front, which may reach the northern half of eastern Canada.

The importance of these air masses is underlined by the fact that the "western" boreal forest is found between the average position of the summer and winter boundaries of arctic air.[4]

Continentality

The degree of continentality can be determined from the study of annual mean temperatures. Using Johansson's formula, a preliminary map of continentality in Canada has been prepared (fig. 1.1).[5] Although values increase from the oceans towards the centre of the continent, the values are not the same on all three oceans. In contrast to the Pacific coast, which has relatively low values (10 per cent), are the arctic shorelands (over 40 per cent) and even the coasts of Newfoundland (30 per cent). The application of the concept of continentality to Canada suggests that the country is divided into two unequal parts on either side of an almost north-south axis joining the Foxe-Hudson-James basins with the Great Lakes. Hudson Bay prevents the displacement of continentality towards the east, as happens in Eurasia (cf. Verkhoyansk, with an index of 100 per cent), and it is western Canada that has the steepest gradient (10 per cent on the Pacific coast, 60 per cent in Manitoba); the values in eastern Canada vary between 30 per cent on the Atlantic coast and 60 per cent in the interior of the Quebec-Labrador peninsula. In short, the continental character of Canada's climate is modest when compared to its vast area. Continentality as such does not seem to prevent human habitation, as Schefferville, Lynn Lake and Thompson indicate. Conversely, however, oceanic influences are inhibited by the mountains of the west coast as well as by the ice covers of the northern and eastern margins and by the general circulation: the oceanic index is therefore low. However, detailed studies have demonstrated the localized effect of water bodies such as the Gulf of St. Lawrence.

4. R. A. Bryson, "Air Masses, Streamlines and the Boreal Forest", *Geographical Bulletin* 8 (1966): 228–269.
   See also R. G. Barry, "Seasonal Location of the Arctic Front over North America", *Geographical Bulletin* 9 (1967): 79–95.
5. D. K. MacKay and F. A. Cook, "A Preliminary Map of Continentality for Canada", *Geographical Bulletin*, No. 20, 1963, pp. 76–81.

Annual Rhythms

Not so long ago, geography textbooks classed as temperate all of Canada that is not arctic. Such an attitude is very far from reality. In fact, it has been suggested that in all of Canada, only a part of southwestern British Columbia can be called temperate; the majority of the country may indeed be considered intemperate. The two features of a moderate temperature regime —an absence of strong seasonal variations and conditions which are rarely either too hot or too cold—scarcely apply to the year as a whole in Canada.

So far as temperature is concerned, the dominant characteristic of the Canadian climate is its range. From one extreme to the other, actual recorded values are separated by 194°F (108°C)* and even monthly differences exceed 100°F (55.6°C). These extremes are scarcely surprising, of course, in view of the size and

* Officially, the coldest temperature of −81°F (−62.8°C) was at Snag (Y.T.); in southern Canada −72°F (−57.8°C) has been recorded, at White River (48°N, on Lake Superior). The maximum is 113°F (45°C), at Midale (Sask.).

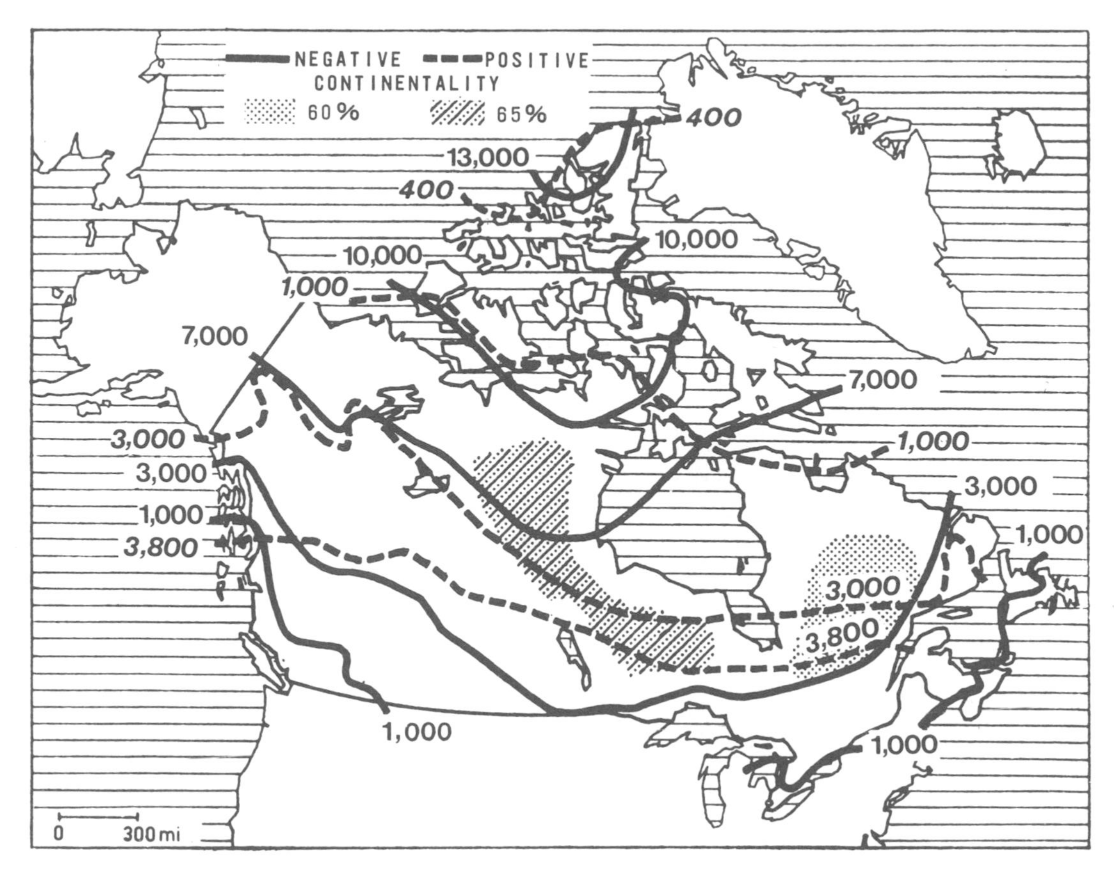

Fig. 1.1 Freezing (negative) and thawing (positive) indices of temperature range. (After H. A. Thompson.)

climatic variability of Canada. Regina, for instance, has recorded 110°F (43.3°C) and −56°F (−48.9°C). The range of variability is best demonstrated by a study of seasonal variations. In table 1.1, we have grouped the *monthly* temperatures in eight simple classes, distinguished by thresholds of which the most geographically significant are: ice, around 32°F; plant growth, 43°F (6.1°C) and 57°F (13.9°C); air conditioning or heating, 65°F (18.3°C). If such calculations had been made on a daily basis, it would have been necessary to include at least two higher classes, hot and very hot.

The thermal diagram (fig. 1.2) clearly shows that each weather station is included in several classes. In the far south of Canada, the temperature varies from warm to cold (*A-E*), while Alert, on Ellesmere Island, experiences temperatures from cool to glacial (*D-G*). Everywhere, the annual range is large: the south experiences months of ice (*E*) and, conversely, the southern Arctic has a near-temperate period (*C*).

To describe periods longer than one month, it is unhelpful to use terms such as *arctic* or *temperate*, since these do not apply to all

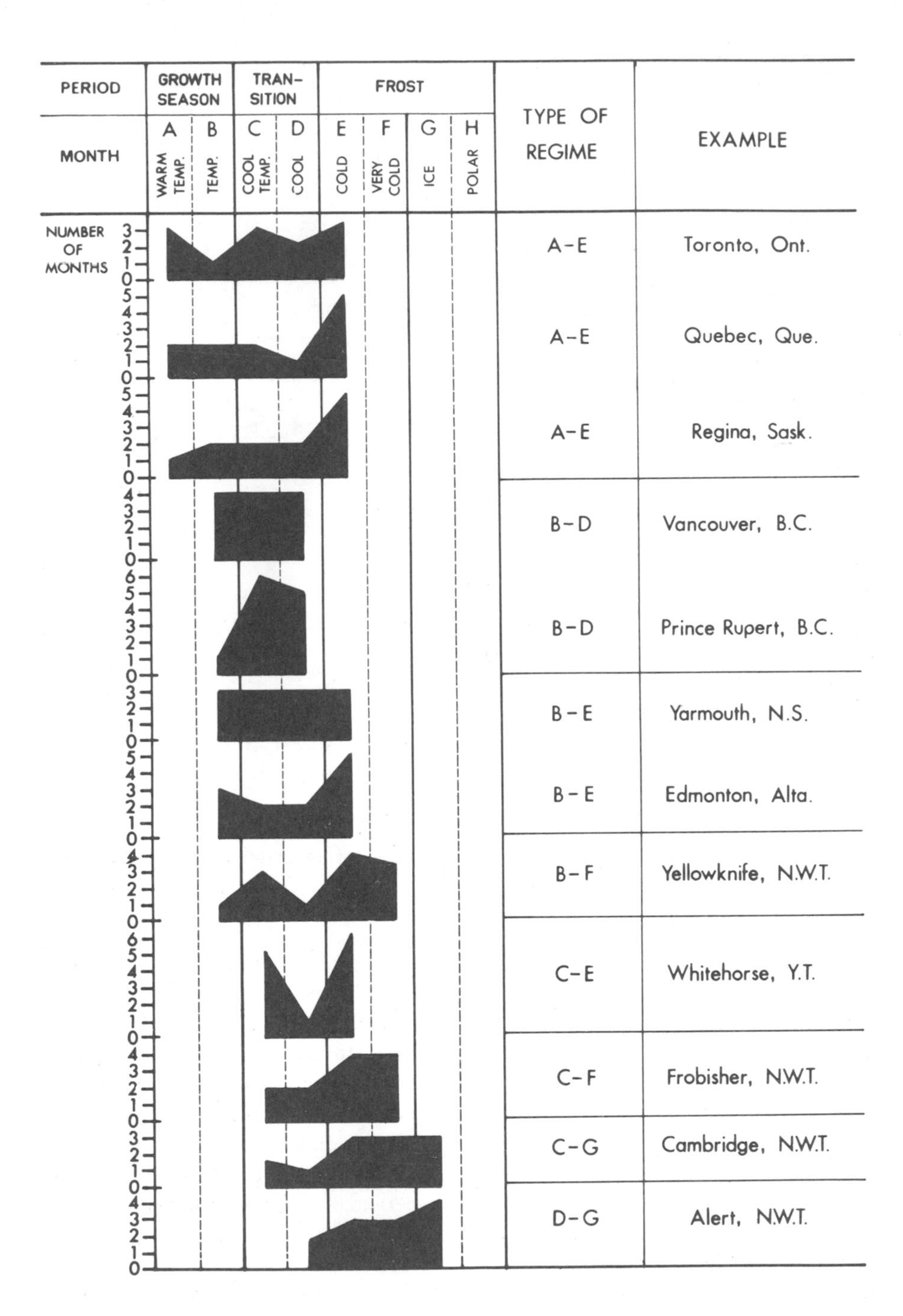

Fig. 1.2 Canadian thermal regimes. (Data from the Canadian Meteorological Service.)

TABLE 1.1 Definition of Temperature Classes

Index Letter	*General Description*	*Limits*	*Major Groups*
A	Warm-Temperate	Above 64°F (18.3°C)	Growing Season
B	Temperate	64°F to 57°F (13.9°C)	
C	Cool-Temperate	56°F to 43°F (6.1°C)	Transitional
D	Cool	42°F to 32°F (0°C)	
E	Cold	31°F to 0°F (−17.8°C)	Freezing Season
F	Very Cold	−1°F to −20°F (−28.9°C)	
G	Glacial	−21°F to −40°F	
H	Polar	Below −40°F (−40°C)	

TABLE 1.2 Monthly Thermal Regime for Toronto

Index Letter	*A*	*B*	*C*	*D*	*E*	*F*	*G*	*H*	*Total (months)*
Months in class	3	1	3	2	3	0	0	0	12
Seasonal type	Growing		Transitional		Freezing				
Months of each type	4		5		3				12

SOURCE: Canadian Meteorological Service.

seasons; the area termed temperate, for example, includes several cold months. Rather, we should describe the annual regime in terms of the letters (from *A* to *G*), or in terms of the locality concerned (e.g., Toronto). The types and subtypes are defined by the range of their monthly values: type *A-E* includes at least one month with a mean temperature over 64°F (*A*) and another in the range 0° to 31°F (*E*). From the number of months in each of the eight classes, and then in each of the three periods, it is possible to establish the temperature profile of a weather station. In the case of the Toronto area, this would appear as 4-5-3 (table 1.2).

Such a temperature range does not define seasons in the strict sense of the word. In southern Canada, instead of the four seasons characteristic of western Europe, there are three periods: the period of growth, transition and freezing (the second of these is not a good time for agriculture). On the basis of monthly temperatures, the northern two-thirds of Canada experiences only freezing and transitional seasons. In addition, the periods are always displaced: the cold period persists into a spring which is "more the death-throes of winter than the dawn of summer" (André Giroux). Because of spatial differences, similar periods do not occur at the same time in the interior of the country (table 1.3).

(1) *Type A-E.* Toronto, Penticton and Regina have at least one month in which the mean monthly temperature is over 64°F (18.3°C). The letter *E* indicates that there are also freezing months, occurring especially at Regina because of the effect of continentality. Penticton is a mountain subtype. Although Toronto and Regina are both *A-E*, the monthly sequences of the two sites are 4-5-3 and 3-4-5, showing a difference in the intra-annual characteristics.

(In terms of length, the Regina situation is characterized more by freezing and less by a period of growth.)

(2) *B-D.* This coastal type is the mildest in Canada. Vancouver (4-8-0) has no freezing month. Prince Rupert, on the margin of the Middle North, is significantly cooler than the shores of the Strait of Georgia.

(3) *B-E.* Maximum monthly temperatures scarcely exceed 60°F (15.6°C); conversely, there are no monthly means below 0°F (−17.8°C). Two subtypes should be distinguished, according to the number of freezing months: a continental type (five months at Calgary) and an Atlantic type (three months at Yarmouth).

(4) *B-F.* This type clearly illustrates the distinction between summer and winter, especially at Fort Simpson on the Mackenzie, which has as many as three months above 57°F (13.9°C) but also has seven months below freezing.

Other typical but less desirable sites do not have any months with a growing season; their first index letter is *C* or *D*.

(5) *C-E.* At Whitehorse the year is evenly divided between transitional months and freezing months.

(6) *C-F and C-G.* Two-thirds of the year show monthly temperatures below freezing. Two subtypes should be distinguished: an interior (Cambridge Bay) and an oceanic (Frobisher), the latter having fewer very cold months than the former.

(7) *D-G.* Here, the largest group of months is in the glacial category (colder than −20°F). Examples are Resolute (0-3-9) and especially Alert (0-2-10).

In each of these major groupings, the seasonal variation is the dominant characteristic. No type has more than four months above 57°F (13.9°C) and all types have cool months 43°F to 32°F (6.1°C to 0°C). Nowhere, however, is there a mean monthly temperature below −40°F (class *H*). Even in this technological era, these temperature variations and the relatively few months of plant growth have a major impact on Canadian life.

TABLE 1.3 Seasonal Types (Number of Months in Each Type for Representative Stations in Canada)

Type	*Station*	*Seasons*		
		Growing	*Transitional*	*Freezing*
A–E	Toronto, Ont.	4	5	3
	Penticton, B.C.	4	5	3
	Halifax, N.S.	4	4	4
	Charlottetown, P.E.I.	4	4	4
	Montreal, Que.	4	4	4
	Ottawa, Ont.	4	4	4
	London, Ont.	4	4	4
	Quebec, Que.	4	3	5
	Winnipeg, Man.	4	3	5
	Chatham, N.B.	3	5	4
	Moncton, N.B.	3	5	4
	Le Bic, Que.	3	4	5

Type	Station	Seasons		
		Growing	Transitional	Freezing
A–E	North Bay, Ont.	3	4	5
	Regina, Sask.	3	4	5
	Saskatoon, Sask.	3	4	5
	Swift Current, Sask.	3	4	5
B–D	Vancouver, B.C.	4	8	0
	Victoria, B.C.	4	8	0
	Prince Rupert, B.C.	1	11	0
B–E	Yarmouth, N.S.	3	6	3
	Sydney, N.S.	3	5	4
	Edmonton, Alta.	3	5	4
	Bagotville, Que.	3	4	5
	Thunder Bay, Ont.	3	4	5
	St. John's, Nfld.	2	6	4
	Gander, Nfld.	2	6	4
	Saint John, N.B.	2	6	4
	Calgary, Alta.	2	5	5
	Grande Prairie, Alta.	2	5	5
	Prince George, B.C.	2	5	5
	Sept-Iles, Que.	2	4	6
B–F	The Pas, Man.	3	4	5
	Fort Nelson, B.C.	3	4	5
	Kapuskasing, Ont.	3	3	6
	Fort Simpson, N.W.T.	3	2	7
	Goose Bay, Nfld.	2	4	6
	Yellowknife, N.W.T.	1	4	7
C–E	Whitehorse, Y.T.	0	6	6
C–F	Schefferville, Que.	0	5	7
	Frobisher, N.W.T.	0	4	8
	Churchill, Man.	0	4	8
	Aklavik, N.W.T.	0	4	8
C–G	Coral Harbour, N.W.T.	0	4	8
	Cambridge Bay, N.W.T.	0	3	9
D–G	Resolute, N.W.T.	0	3	9
	Alert, N.W.T.	0	2	10

SOURCE: Canadian Meteorological Service.

Plate 1.1 Stunted black spruce close to the treeline at Inuvik, N.W.T. Dwarf tree species are found far beyond the conventional treeline, and the arctic willow (*Salix arctica*) extends as far as the northern limit of land on Ellesmere Island, N.W.T.

Freeze-Thaw Oscillations

One of the most important climatic thresholds is freezing point. In a variety of ways, the conversion of water from the liquid to the solid state and vice versa is of major importance: these freeze-thaw oscillations shatter rocks, arrest plant growth, and affect road conditions and navigable waterways. In Canada, cold-weather problems are not caused by the winter as a whole so much as by the oscillations of temperature around 32°F; these oscillations are as inconvenient as they are numerous in the inhabited areas of Canada and they can occur at the most unexpected times, in very late spring or very early fall. The winter of 1949–50, for example, was catastrophic for the orchards of British Columbia. Such events tend to curtail still further the really "fine" season.

In terms of thresholds at 28°F (−2.2°C) and 34°F (1.1°C), the total number of annual freeze-thaw oscillations is generally 50 to 60 in the nonmountainous parts of southern Canada, 25 to 40 around Hudson Bay and 15 to 30 in the Arctic Archipelago. In other words, there is a decrease from south to north. There are local variations, as at Franz (Ontario), where the effect of Lake Superior causes frequencies to be as high as 76. The Cordillera is marked by extremes: on the coastal side there are scarcely 10 oscillations a year; in some mountainous areas there are more than 100.

Similar contrasts exist between north and south in the occurrence of such oscillations during the year. In the High Arctic, they are restricted to a very short period on either side of true winter conditions; in southern Canada, by contrast, they occur throughout half the year and, because of the frequent milder spells, they may take place in the middle of the coldest period. Moreover, the *isopalimpex* (a line joining places with an equal number of cycles) of J. K. Fraser shows that oscillations are more common on the coast than in the interior, maritime air being more unstable.[6] In the Far North, the limited number of cycles, together with the low humidity, causes problems for geomorphologists when contrasted to the vast areas of frost-shattered rocks.

Plants are sensitive to ice and to the recurrence of premature onset of frost during the periods of growth and fruiting. According to the *Atlas of Canada*, the average length of time between the last daily temperature below 32°F (0°C) in spring and the first in fall is about 120 days in the Laurentian Valley, 180 in the favoured parts of Ontario and 100 on the Prairies.[7] However, it is not sufficient to take into consideration only the threshold of 32°F, nor to look only at averages; plants such as potatoes can tolerate low temperatures. Liability to frost varies according to topography, the stage of growth, soil character, plant type and atmospheric conditions. Agronomists have preferred to take extreme thresholds at 20°F (−6.7°C) and 40°F (4.4°C). In the Laurentian Valley, the frost-free period (with an 80 per cent probability) is 45 days if the threshold is 40°F, 102 days if it is 32°F, and 181 days if it is 20°F. Climatic conditions are extremely variable. Such wide variation around the freezing threshold shortens the dependable season, and encourages neither intensive agriculture nor the growth of delicate plants. The liability to frost, by favouring resistant species, has reduced the number of forest species and cultivable plants, but has also encouraged research stations to develop rapidly maturing species. As the zonation would lead one to expect, the growing season becomes progressively shorter and less dependable from south to north (fig. 1.3).

Human Response to Climate

People differ in their perception of weather and climate. Various researchers have calculated the critical thresholds of moist air and especially of oppressive hot and humid weather. These depend on an inverse relationship between temperature and relative humidity. In terms of comfort the effect varies by season. In continental regions during winter, except in depression weather, the air is generally dry; this is fortunate, since humidity would greatly accentuate the physiological effects of the cold. Humidity is mainly a nuisance during the summer; in the St. Lawrence Lowlands the development and the stagnation of warm and moist air masses coming from the south and southeast gives a sultry atmosphere. In places where the actual temperature exceeds 100°F (38.7°C), the weather can become as hot as in the tropics, especially if there is a cloud cover. (Aklavik, although north of the Arctic Circle, has recorded 93°F [33.9°C].) Urban housing, which is poorly protected from incoming solar radiation and lacks the ability to remove warm and moist air rapidly, is very uncomfortable during the two or three moist heat waves of summer. In summer, Toronto is subjected to tropical air masses lasting about four days on an average of about four occasions; during the months of July and August, discomfort would be felt by a large majority of the population.[8]

---

6. J. K. Fraser, "Freeze-Thaw Frequencies and Mechanical Weathering in Canada", *Arctic* 12 (1959): 40–53.
7. [Canada], *Atlas of Canada* (Ottawa, 1957), plate 23.
8. D. Kerr and J. Spelt, *The Changing Face of Toronto*, p. 23.

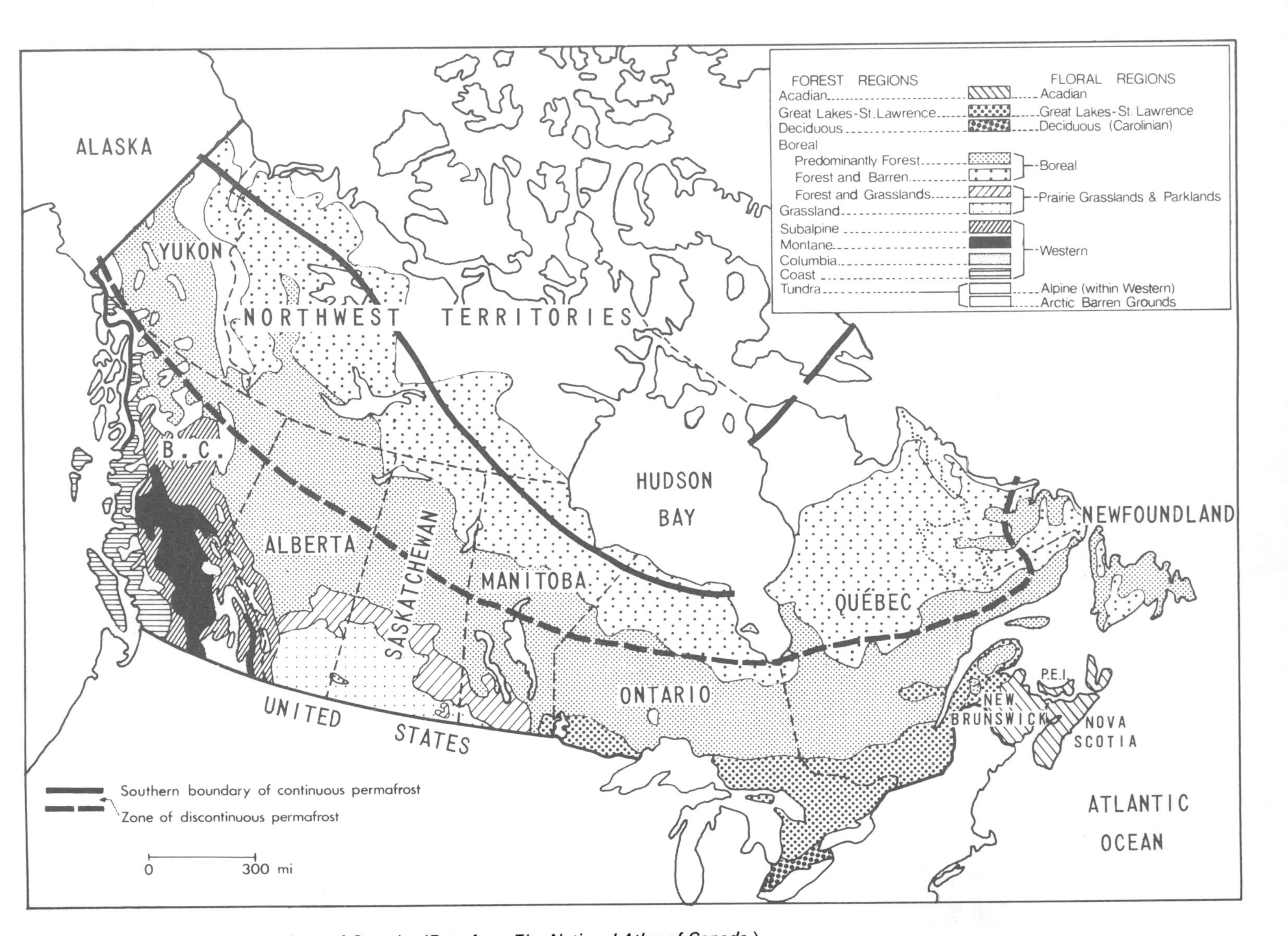

Fig. 1.3 Forest and floral regions of Canada. (Data from *The National Atlas of Canada.*)

In Montreal, during the summer months, fresh and windy days represent 39 per cent of the weather; pleasant warmth and humidity, 28 per cent; rainfall, 20 per cent; and very hot days, 13 per cent. Days in which it feels good to be alive are neither regular nor numerous. Unpleasant weather tends to send the city dweller into the country. In offices, air-conditioning is necessary. Such solutions are not always adequate and, paradoxically, the settled portions of Canada may perhaps suffer more from heat than from cold.

In the High Arctic, the extremely dry air leads to a certain dehydration, while the effect of what moisture exists is to cover everything—clothes, faces, houses, boats and wires—with hoar frost.

Like humidity, wind increases the cooling effect. A temperature of 5°F (−15°C) with a wind of 25 mph has been shown to have a cooling effect on the skin equivalent to that of a −40°F (−40°C) temperature in calm conditions; hence, because of the wind systems, the rigours of arctic conditions are not restricted only to polar areas. According to Siple's index, most of the country experiences, in January, more than 1,100 windchill units, a threshold above which it is termed "very cold" (in KCal/m^2/hr). The Canadian maximum of 1,900 does not occur in the Arctic Archipelago but in continental Keewatin. Windchill is hence distributed neither latitudinally nor by major zones. Winnipeg, Manitoba, at 50°N, has a higher average index than has Dawson, in the Yukon, at 64°N.

Physiological cold—loss of heat from a human being—leads to either local or general hypothermia, depending on whether it affects an individual limb or, as in serious cases, the circulatory system.

There are various methods of eliminating this heat loss. It is possible to develop a resistance to cold by adjusting to it mentally and physically, the latter of which is aided by a sufficient calorie content in the diet. In traditional Eskimo life the oil lamp in the igloo contributed to thermal equilibrium. Physical activity itself reduces the demand for artificial heat; at 32°F (0°C) one would need a single "clothing unit"* while taking part in strenuous physical activity, but four units while at rest.[9] The amount of clothing depends on a variety of factors and there is an enormous difference between summer conditions indoors in southern Canada and a winter storm in Keewatin; in clothing units the extreme range is probably from less than one to more than ten. In southern Canada, "Canadians have conquered winter";[10] artificial heating permits one to work indoors, in comfortably light clothing. Indeed, modern life in cities has almost confined contact with the real outdoor climate to recreational activities.

A few decades ago, the climate studied by geographers was essentially that related to agricultural requirements; today, there is also a concern for urban climate, in which the vast majority of people live. Concentration of population has created artificial pollution through vehicles, power and heating, industrial activities and waste disposal. "The atmosphere of all modern cities is pervaded by the continuous presence of various contaminants";[11] respiratory diseases are among the effects of such pollution. In the last few years, the inhabitants of metropolitan areas have developed an acute awareness of air pollution.

---

* The amount of clothing necessary for a resting person to be comfortable in a room heated to 72°F (22.2°C), where the humidity does not exceed 50 per cent.

9. J. E. Sater, ed., "Living in the Arctic", *The Arctic Basin* (Centerville: Tidewater Publication, 1963), pp. 278–288.
10. See P. Deffontaines, *L'homme et l'hiver au Canada.*
11. A. Auliciems and I. Burton, "Air Pollution in Toronto", in W. R. Sewell and I. Burton, *Perceptions and Attitudes in Resources Management*, Resource Paper, No. 2 (Ottawa: Information Canada, Department of Energy, Mines and Resources, 1971), p. 74.

Chapter 2

Frozen Ground

In northern countries cold affects all aspects of the geographical environment, not merely air and water but also the surface layers of the ground. *Geocryology*, the study of permafrost, has suffered from chronic neglect, yet everywhere in Canada the ground freezes and on Melville Island in the High Arctic the permafrost layer extends 1,800 feet (548 metres) below the surface. Frozen ground can serve as an indicator both of past and present temperature conditions and of seasonal and regional climatic differences.

Permafrost

Pergelisol, *permafrost*, *permagel* and *cryolith* are all terms used to describe ground which has been frozen permanently—at least for hundreds of years. The ice that binds together the ground material (rock, unconsolidated material or organic matter) may be abundant (*humid permafrost*) or scarce (*dry permafrost*). The temperature varies from about 14°F (−10°C) to about 32°F (0°C); sections in both vertical and horizontal planes show marked differences from place to place and may include unfrozen portions (*talik*). The upper and lower boundaries of the permafrost are not perfectly level but vary according to topography, moisture and positive or negative heat transfers. Where glaciers or water bodies extend over a wide area, they tend to insulate the ground and prevent the deep penetration of atmospheric freezing.

Ideally, a vertical section through permafrost shows a series of zones in two main bands, the *active layer* and the *true permafrost*.

The Active Layer

This thin layer usually melts each summer. In terms of thermal behaviour there is a significant difference between its upper surface, which, depending on insolation receipts, may pass through several freeze-thaw oscillations each year, and the section below this, several inches thick, which is much less sensitive to such temporary changes. The overall depth of the active layer varies from one year to the next, from one soil type to another and also from north to south in Canada. In the Arnaud (Payne) basin of arctic Ungava, the author, in August 1955, measured a depth of 12 inches (30 centimetres) in organic terrain, 29 inches (75 centimetres) in a boulder field and 4 feet (1.22 metres) in well-drained gravel terraces. Near Chimo, Quebec, maximum thawing of the active layer lasts for four or five months, ending in the deepest part at the end of October, by which time the freezing of the surface layers has already begun.[1] At any given time there is a difference between the air temperature at 4 feet (1.22 metres) above the ground and the temperature at different depths in the active layer.

True Permafrost

It seems that in the High Arctic, the deeper layers—the true permafrost—consist of two zones. The first, the upper zone, measures around 100 feet in depth, and is affected by a marked loss of heat. The top of this zone is the upper boundary of the true permafrost ("permafrost table"). The second, or lower zone, becomes much less cold with increasing depth. The irregular base of this layer constitutes the lower boundary of the permafrost, which at Resolute is found at about 1,300 feet (400 metres).

---

1. See G. Robitaille, *Observations sur le dégel saisonnier. Région de Fort-Chimo (Québec)*, Vol. 15 (Quebec: Centre d'Études Nordiques, 1967).

Plate 2.1 A length of the experimental pipeline built near Inuvik, N.W.T., by Mackenzie Valley Pipeline Research Ltd. Because oil must be conveyed at high temperatures, an underground pipeline would cause melting of the active layer of the permafrost, leading to soil movement and the risk of pipeline fracture and spillage. A surface pipeline, as illustrated, has its own problems: it impedes surface drainage in summer and may represent a barrier to migrating wildlife.

Subpermafrost. Lower still, where the ground temperature is above 32°F (0°C), is the subpermafrost.

In spatial terms, permafrost extends over the northern half of Canada, in two types of distribution. In the Far North, permafrost is *continuous*, whereas in the high Subarctic it is *discontinuous*, since ice-free areas separate frozen areas. The scattered permafrost conditions of central Quebec-Labrador provide an example of the latter distribution. There is a significant latitudinal difference between the Labrador coast, where the southern boundary occurs at about 55°N, and the southern Yukon, where it retreats to 60°N. It is on the southwest side of Hudson Bay that discontinuous permafrost reaches farthest south, to about 52°N. South of the discontinuous permafrost, occasional (*sporadic*) permafrost occurs; that found in the Upper Gaspé belongs in this category.[2]

The geographical effects of permafrost are obvious. The upper surface of the frozen ground is a barrier to the penetration of water and to the spread of plant roots. Frozen ground also affects mining conditions; iron ore mining

2. See R. J. E. Brown, *Permafrost in Canada; Its Influence on Northern Development.*

in Labrador is hindered by lenses of permafrost. Urban settlements in the Arctic have been forced to distribute water and collect sewage by means of surface "utilidors", a form of heated and insulated conduit. The active layer is, however, the greatest nuisance. Roads, trails, buildings and even sewage beds must be carefully located so as to minimize disturbance of natural conditions and avoid impounding water which might subsequently cause any number of problems through ice formation. Such conditions have led to the design of special types of houses. Many of the workers at arctic radar sites, for example, live in bunkhouses which are long and narrow and which are grounded in surface deposits unlikely to contain ice; a layer of air also insulates the floor of the building from the surface of the ground. Another method is to freeze the active layer in summer.

The whole range of periglacial landforms, especially patterned ground, is intimately related to the occurrence of seasonal active layers and permafrost.

Ground Ice

In most cases a distinction can be made between the small amount of ice which binds together frozen ground and local masses of almost pure ice. Such ice cores are themselves of various types; they include the cores of *pingos*, the conical hillocks of which more than a thousand occur on what was once the delta of a proto-Mackenzie.[3] They include also the small *fissures* or *wedges* of ice which often define the edges of polygons on the tundra. A third type consists of large horizontal sheets of ice which develop in areas where considerable underground movement of water has occurred in a fairly deep active layer. In organic terrain, small ice *lenses* have been found as far south as 48°N (Abitibi). When exposed to the air, these ice cores melt rapidly, creating thermokarst depressions or leading to landslips. Such landscape evolution might be termed *fluviothermal* if it is associated with stream action.

3. See J. R. Mackay, *Mackenzie Delta Area.*

Temporarily Frozen Ground in Southern Canada

Frozen ground has been studied far less widely than air temperature. The types of frozen ground are very variable; for example, snow or plant covers act as insulators and affect the freezing of soil moisture, which is controlled by cold air. If snowfall occurs at the onset of winter, the ground beneath scarcely freezes; when winter snowfall arrives late, however, the cold wave may penetrate the ground to a depth of several feet and roots and underground pipes may fracture.

Ground thaw comes only after the surface snow has disappeared. This means that the rise in ground temperature in spring is delayed in comparison to the rise in air temperature. When this happens problems for tree growth result, since the branches need to transpire but the roots cannot yet deliver the necessary quantity of moisture. This may be the reason for a massive decline of the birch in eastern Canada. Percolation of surface water is necessarily very small and hence the recharge of soil moisture does not reach its potential.

Ground melting produces phenomena such as outpourings of mud similar to those of an arctic active layer. Movements of material are apparent even on asphalt surfaces as minor indentations are deepened by wheels into potholes and road margins disintegrate rapidly. To reduce such damage, heavy trucks are banned for several weeks in springtime. Building the road on a special, well-drained embankment, similar to that of a railway track, lessens such problems. In cultivated fields, freezing of the soil causes a form of climatic fallowing. Frozen ground also affects housing, especially simple structures which have no foundations, whence the rapid deterioration of barns, porches and wooden garages.

Organic Terrain

The development of organic terrain is a function of cold conditions, either past or present. This is a matter of some importance, since approximately 500,000 square miles (1.3 million square kilometres), or 14 per cent of the land area of Canada, is covered with this vegetation type, collectively termed *bog* or *muskeg*. Organic terrain is found throughout the southern half of Canada, since this area is more suitable than the Arctic for the growth of semiaquatic plants. The lowlands south of Hudson Bay, the interior plateau of Labrador, the Mackenzie basin and coastal regions which have been submerged in recent geological times are principal areas.

There is a great variety of muskeg types. Seen from above, flat and featureless muskegs do not form the majority; most show some form of regular arrangement of surface features. The surface of the bogs may be composed of water areas separated by strings of vegetation; these "flarks" are generally formed in parallel lines spaced various distances apart. If they are straight, or almost so, they are described as having a rectilinear pattern; if curved, they are termed *concentric*. Some formations are more *dendritic* (tree-shaped) in appearance; other less geometrical formations have a vermicular, wavy appearance. Some branched patterns of strings form a strict braided pattern, while others with discontinuous arrangements of vegetation sometimes form circular or polygonal clumps, described as *checkered*. Seen from above, the bog may contain depressions caused by melting ice or, in contrast, it may be ridged with hummocks of frozen material called *palsen*. Normally these patterns do not quite reach the outer edge of the bogs. The various arrangements of plant material reflect, to some extent, the influence of periglacial factors.[4]

Although there are many overlaps, a chronological sequence seems to extend from the north-central parts of Canada towards the south. Some active bogs are extending in the low Arctic and many more in the Subarctic, whereas "islands of vegetation" and forested strips are found in the muskegs to the south of James Bay. In southern Canada many of the bogs have become forested while others, less covered with trees, show traces of the former strips.

Until very recently these areas of muskeg and organic soil had been avoided for agricultural settlement and communication routes. However, some bogs (for example, those along the St. Lawrence estuary) have become sites of mining activity and vehicles have been devised for travelling over muskeg areas.

Frozen ground is of outstanding importance. Nearly 50 per cent of the land area of Canada lies within the zones of continuous or discontinuous permafrost. For part of the year, the entire surface is frozen. For between one and three months, during the thaw, the ground is soft and unworkable. Even though the melting of temporarily frozen ground in the south of Canada differs in characteristics from that occurring in the active layer of the Arctic, the damage which is caused, especially to communications, affects everyone; repairs are part of the total cost of the winter. Revenue and

---

4. N. W. Radforth, "Organic Terrain and Geomorphology", *The Canadian Geographer* 6 (1962): 166–171.
See also L.-E. Hamelin, "De Winnipeg au Keewatin", *Revue de Géographie de Montréal* (1971): 89–94.

taxes must be used to restore the situation, though such investments are essentially unproductive. Voltaire, who deeply wounded the inhabitants of Canada by describing their country as "a most detestable land of the North", was not entirely incorrect.

Chapter 3

A Geography of Water

Precipitation

Differences in precipitation distinguish the various Canadian environments from one another. In terms of origin, rainfall may be derived from frontal movements; it may be related to areas of significant relief or it may come from convection and turbulence mechanisms. Most precipitation is derived from the first two causes, which in turn depend on the behaviour of the general circulation. It has been noted that without the moisture which is derived from the Gulf of Mexico, the total precipitation in southern Canada east of the Rockies would be as low as that of central Asia.

Depressions and anticyclones are normal components of the tropospheric circulation. During winter the polar anticyclone that occupies the northern part of the continent is fringed by two areas of depressions, in the Gulf of Alaska and southeast Greenland. In summer, by contrast, the continental anticyclone retreats northward and weakens, while the low-pressure areas over the Atlantic and the North Pacific become more significant. The continent is also invaded by travelling depressions which differ little in their trajectories from summer to winter. One series moves from Alaska in the direction of central North America and then gradually alters course to reach the Atlantic in the vicinity of Newfoundland; another set, moving from south to east, often reaches the St. Lawrence Lowlands. As well as these large moving systems, more stationary air masses also occupy various parts of the continent.

During the cold period, precipitation is abundant on the Pacific coast and moderate in the Atlantic Provinces and in these areas is related to the nearby oceanic low-pressure systems. Over the majority of the country, however, summer is a slightly wetter season. This moderate maximum results from the more frequent northern penetration of humid air; thus there is little evidence of continentality in precipitation. However, in the Prairies and the Far North, precipitation occurs mostly in the summer.

On the national scale, the spatial distribution of precipitation may be summarized as follows:

(1) Only one area, the narrow band along the coast of the Pacific, is markedly wet, with about 100 inches (or 254 centimetres) per year. North Vancouver, at 1,130 feet (344.4 metres) above sea level, receives 102 inches (259.39 centimetres). This abundant rainfall leads to the most vigorous forest growth in Canada.

(2) The St. Lawrence axis, traversed regularly by depressions, gets 30 to 40 inches (76 to 100 centimetres) a year. Despite high evaporation in summer, this has allowed agriculture to become a factor in the spread of population and economic activity.

(3) By contrast, separated from the Pacific by the Cordillera and off the track of movements from the Gulf of Mexico, the greater part of the agricultural ecumene—the southern part of the Prairie Provinces—only has 12 to 18 inches (30 to 45 centimetres) a year.

This is the more serious because the annual variability is about 25 per cent. The effects of lack of moisture on agriculture were severe for several years after 1929.

(4) In the interior of the Rockies topographic conditions give rise to arid situations. Ashcroft, on the Thompson River, only receives an average of 7.5 inches (19.18 centimetres) per year. Irrigation is therefore essential in the various longitudinal valleys of the Columbia system.

(5) In the Middle North, the probability of frontal rain decreases and precipitation progressively lessens. This northward gradient is particularly significant in the Yukon.

(6) The High Arctic is a dry area on the precipitation map. Alert and Resolute get the equivalent of about 5 inches (13 centimetres) of water and the Parry Island arc, still less. The Lake Hazen valley on Ellesmere Island averages only 1 inch (2.5 centimetres) of precipitation a year, making it virtually a desert. Although the Far North has little rainfall, it should not be concluded that in August the ground surface is dry; the surface moisture produced by melting of the active layer of permafrost gives more the impression of a humid landscape. Comparison between hot deserts and such cold deserts should therefore not be pushed too far. Farther south, on the edge of the Arctic, precipitation is insufficient for plant growth (both the lack

Plate 3.1 Summertime in Canon Fiord, Ellesmere Island. A few small icebergs float in the sea after calving from a nearby glacier. During the two months or so of above-freezing temperatures, geomorphological processes may operate with dramatic speed in a landscape that has little plant cover. Note the evidence of slumping, the active delta formation and the types of slope development. At this season, the main restriction on geomorphological activity is the very low summer rainfall: this is a "cold desert".

of rainfall and the cold limiting vegetation here).

Snow Everywhere

Snow is one of the most characteristic seasonal features of the Canadian landscape. A test conducted among young adults showed that snow is more an identifying feature of winter than cold, floating ice or wind.

The geography of snow begins with a study of winter falls. In non-Cordilleran Canada, such falls are relatively limited since winter is a comparatively dry season in the annual cycle. The proportion of snow to total annual precipitation is low; Toronto records only about 5.4 inches of water derived from snow out of an annual total of 30.9 inches. In the continental interior—at Churchill and Kapuskasing, for example—the proportion falling as snow reaches one-third. Only on the edge of the Arctic—at Cambridge, Coral Harbour, Frobisher or Yellowknife, and even at Resolute—is the ratio around 50 per cent. At Aklavik, snow accounts for 55 per cent; at Alert, between the Arctic Ocean and the "Innuitian Alps", it sometimes reaches 80 per cent. Over the greater part of Canada total precipitation is therefore dominated by rainfall. Rivers and streams derive more of their flow from rain than from melting snow.

Even though the proportion which falls as snow increases northwards, the actual amount of snow decreases markedly. Montreal receives 100 inches (254 centimetres) of snow a year (10 inches of water-equivalent), whereas Churchill has only 50 inches (127 centimetres) and Resolute, less than 28 inches (71 centimetres). Any idea that northern Canada is very snowy must therefore be dismissed. Some stations, it is true, do receive large amounts: in the southeast of the Quebec-Labrador peninsula higher sites record 200 inches (508 centimetres). The Canadian maximum is determined by altitude, by closeness to the sea and by the liability of winter precipitation. At 5,000 to 6,500 feet (1,500 to 2,000 metres) slopes open to the Pacific receive approximately 50 feet (15.25 metres) of snow, with a maximum of around 75 feet. Such conditions have been known to lead to avalanches.

Snowfall normally occurs at temperatures close to 32°F (0°C). Over Canada as a whole, though the number of days per year with snow does not often exceed sixty, their distribution during the year varies from one region to another. In the south, snow is entirely a winter phenomenon; in the Middle North, days with snowfall tend to occur mostly in spring and autumn. In the Far North, snow can fall in each month of the year, the minimum frequency occurring in the middle of the winter anticyclone.

Neither the low winter snowfalls of the Arctic nor the summer warmth in the southern mountains assists the development of large or thick glaciers. Almost all the ice-covered areas are to be found in the coastal mountain chains of the Yukon and British Columbia in the west or those on Baffin and Ellesmere islands in the east.

By themselves, days with snowfall do not serve to define the cold season since they are only one of the four components of winter climate. The first type includes days of storms, wind, poor visibility, snow on the ground, and consequent delays in communications (roads temporarily blocked), and absences from work and church. On 11 and 12 December 1944, Toronto experienced a severe winter storm which brought 20.5 inches (52 centimetres) of snow; and in March 1972, Quebec City was snowed under for a week after the worst snowstorm in its history. The second main type of winter weather is seen in the days of severe cold, with dry air, strong light and echoes; this weather encourages lively activity and winter sports. In the Laurentians, these two types of weather alternate for about four months, in a rhythm of about one day of storm for each five or six days of clear or changeable weather. This pattern is often disturbed, since variability is the main characteristic of the climate. At Quebec City the winter, which generally lasts from the end of November to the end of March, includes four alternating types of weather. Storms characterize 10 to 20 per cent of winter days, severe cold 5 to 15 per cent, periods of spring-like weather about 25 per cent and the rest, characterized by moderate cold, represents transitions between the three preceding categories.

Although there are four structural elements, we can distinguish between *prewinter* in November and December (little sun, little snow, freezing rain and sleet), *midwinter* (storms, snow cover, cold, sunshine) and *winter's end* (light, meltwater).

The Snow Cover

A clear distinction must be made between the amount of snow which falls and the amount which accumulates on the ground. Snow is a fragile material, moveable and capable of compaction; what remains on the ground may therefore be only a small fraction of what falls. At the end of the main winter period, for example, Quebec City, which has received about 123 inches (312.4 centimetres) of snow, has barely 25 inches (63.5 centimetres); during the cold season in southern Canada the snow on the ground compacts and its density can exceed 0.3.

The depth of the snow cover is greater in the interior of the country than in the southeast (Nova Scotia), southwest (Vancouver), south (Ontario peninsula) or even the north (High Arctic); as with temperature and total precipitation, snow cover conditions vary considerably from place to place. In eastern Canada, because occasional periods of local melt occur even in the middle of the cold season, the snow cover is relatively thin when compared to the heavy precipitation. Refreezing of water in the snow creates hard layers or crusts within it. Little snow accumulates on the ground in the Arctic since the wind redistributes it into depressions or against barriers; even in midwinter the landscape of the Far North is not as freshly white as that of the Quebec-Labrador peninsula.

The duration of a snow cover, defined as a thickness of at least 1 inch (2.5 centimetres), varies more regularly across the country. There is a progressive increase from south to north, from 60 days on Lake Erie to 300 in northern Ellesmere Island.[1] Here again, however, variation from year to year reduces the significance of these average values.

Snow sharply distinguishes the Canadian environment. It is there each year and everywhere, completely altering the colours of the landscape. In May, before the renewal of plant growth, southern Canada appears very dull by comparison with February and March, when the snow-covered forests take on a bluish, nostalgic appearance (which Canadian painters have sought to capture). There are, then, actually two different Canadas: that of the nonsnow period and that of winter. In the settled portions of the country, snowfall used to put an end to movement, and even today, only the train is able to maintain a regular schedule across long distances on stormy days (e.g., 17 December 1969 in the St. Lawrence Lowlands). Snow is expensive: snow removal in southern Quebec costs the Ministry of Highways about $30 million a year. The removal of snow from the roads also allows freezing temperatures to penetrate more deeply, which in turn leads to considerable damage at the time of melt.

Snow, by covering ground and rock outcrops, restricts agriculture and some mining activity. On the other hand, the forest industry and some areas of farming benefit, as it is easier to establish temporary roads on the snow than on the ground. Tractor convoys that penetrate far into the North in winter use both the snow cover and the ice cover on water bodies. The prolonged

1. See J. G. Potter, *Snow Cover*, Dept. of Transport Climatological Studies, No. 3 (Toronto, 1965).

snow season assures skiers a worthwhile season, especially in the Laurentians north of Montreal and in the Cordillera. Winter snow affects rivers and streams, as melting leads to run-off which is abundant by contrast with the preceding period of ice. Meltwater is also of major significance for vegetation at the start of the growth season.

A snow cover, especially of fresh snow, dazzles the observer. In the Arctic, in springtime, a diffuse light without shadows that sometimes extends over a vast area may cause the observer to lose his perception of distance (*whiteout*). To prevent aircraft pilots from being victims of this snow dazzle, the margins of small airstrips are marked by the dark colours of empty oil barrels, enabling better judgment of distance.

Floating Ice

Since 1958, the International Geophysical Year, glaciology has included the study of all kinds of ice, including moving ice, and the term *glacial* has been used to refer to the geography dealing with this phenomenon. Moving or floating ice is found not only in high latitudes but throughout Canada. Its annual cycle consists of three phases.

(1) *Freeze-up* takes place by spontaneous freezing, by the inclusion of a water layer between separate ice layers, by mechanical or climatic agglomeration of individual floes, by calving (icebergs) or by direct conversion of snow.

(2) *Ice maximum* occurs either as a complete cover or in the form of a discontinuous cover that consists of marginal or moving (*pack*) ice.

(3) *Break-up* is one of numerous methods of ice disappearance. The period without ice forms an interglacial. Because of weather changes, neither freeze-up nor break-up is a progressive phenomenon; delays and reversals in the process are frequent. Similarly, there is no more coincidence in the dates of snowfall, freeze-up and the freezing of the ground than there is between snowmelt, break-up and the thawing of the ground.

Break-up takes place more slowly in lakes than in rivers; this is not necessarily true of freeze-up. The extreme south of Canada is only locked by ice for about 100 days, but the situation deteriorates rapidly with increasing latitude; away from central Ontario the ice cover may last more than 6 months. Around Axel Heiberg Island it remains for about 330 days. Ice does not follow the pattern of the four astronomical seasons and almost nowhere in Canada is it limited solely to winter. It spreads over into the transitional seasons, but more into spring than into fall.

Floating ice is a universal element of the landscape and its geographical significance has been insufficiently appreciated. Hydrologically the frazils and small floes affect discharge conditions: the ice causes frequent barriers that result in flooding upstream and, if the barrier bursts suddenly, similar catastrophes downstream. Major floods, such as those of the St. Lawrence in 1896 and the Mackenzie in 1961, are often attributed to snow when they are in fact due to ice. Flood conditions can occur at any time in the cold season in southern Canada.

The ice is a significant agent of landscape evolution as well.[2] Along the banks and in the riverbed the ice scours, polishes and striates the bedrock and carries away sedimentary material. The base of ice that develops in the course of a single winter can create a coastal erosion platform. Floating ice is also a powerful agent of transport and accumulation. Millions of transported blocks lie along the shores of the St. Lawrence estuary, along the Arnaud (Payne) River of Ungava and along other streams. On

2. See J. C. Dionne, *Aspects morpho-sedimentologiques du glaciel, en particulier des côtes du Saint-Laurent.*

shorelines the grounding of ice due to the wind, to refreezing or to a rise in water level can bulldoze earth, vegetation and even buildings; along the southern shore of Melville Island in the Arctic a pushed-up ramp 1 mile long and more than 10 feet (3 metres) high shows the effect of the pack ice in Parry Strait. These are merely a few characteristic effects of moving ice.

Ice has contrasting effects on movement. Along the St. Lawrence, the strong shore ice has served as a means of communication; along the arctic margin, tractors prefer the frozen surface of lakes. In the forest exploitation zone, timber is transported (at no cost) during break-up and dumped along this ice cover. If thick enough or artificially strengthened, ice surfaces can be used as landing strips. It is virtually essential to have icebreakers or submarines in order to navigate through or under seas which are heavily ice-covered. Two indices are used to aid ice navigation (often improperly called winter navigation). The *coefficient of cover*, or index of ice concentration, expresses in tenths the proportion of the water surface covered by ice; four- to six-tenths represents a barrier to the majority of ships. The *index of fragmentation*, composed of three digits, provides a measure of the size of floating ice fragments. Thus, for a water body containing no icebergs, a value of 721 would indicate that 70 per cent of the ice surface is composed of floes less than 10 metres (32.5 feet) in diameter, 20 per cent is from 10 to 100 metres and 10 per cent more than 100 metres. In 1964, Mackenzie River navigation began on 10 June at Lake Athabaska, 20 June at Great Slave Lake and in mid-July at Tuktoyaktuk on the arctic coast. The St. Lawrence Seaway is closed to navigation for almost four months.

When a cover of floating ice is established, the surfaces of water areas are no longer visible; instead there is a continuous hard surface from the Arctic Ocean to the U.S.A. Canada consequently appears much more continental in winter than in summer.

Moisture and Plant Growth

Do the natural vegetation and cultivated crops in Canada have enough water? Is the growth of species restricted by lack of water? In particular, is the growing season fairly moist? In order to provide an answer to such fundamental questions, we would have to distinguish among the needs of forty agricultural regions in Canada.

The surpluses and deficits of moisture have been determined through the use of Thornthwaite's potential evapotranspiration index. These calculations indicate, first of all, that eastern Canada and the Prairie Provinces do not share the same characteristics. East of Hudson Bay and Lake Superior there is generally a water surplus. This may exceed 25 inches (63.5 centimetres) in Nova Scotia; in Grand Falls, Newfoundland, too, the amount of moisture available to the forests has been described as "very good". The average annual evaporation at Montreal is, however, about 5,000 Livingston units.

Such calculations also show strong seasonal variations; in the St. Lawrence Lowlands, in relation to plant needs, the water cycle has four components. Periods of surplus occur in winter when evaporation and run-off are small; spring is a period of abundance during which run-off is high; in summer the losses by evaporation are large, while autumn is a time for the recharge of reserves. Periods of surplus (or of sufficient soil moisture) and periods of deficit thus occur side by side. During the Montreal summer, reserves drop gradually, despite the seasonal precipitation peak; summer deficits can even cause concern, as in 1941. In eastern Ontario and western Quebec, a period of relative dryness in 1965 lasted from mid-May to the beginning of July due to an abnormal movement of depressions towards the centre of the continent (they normally pass along the St. Lawrence Lowlands). Most of the Ontario peninsula,

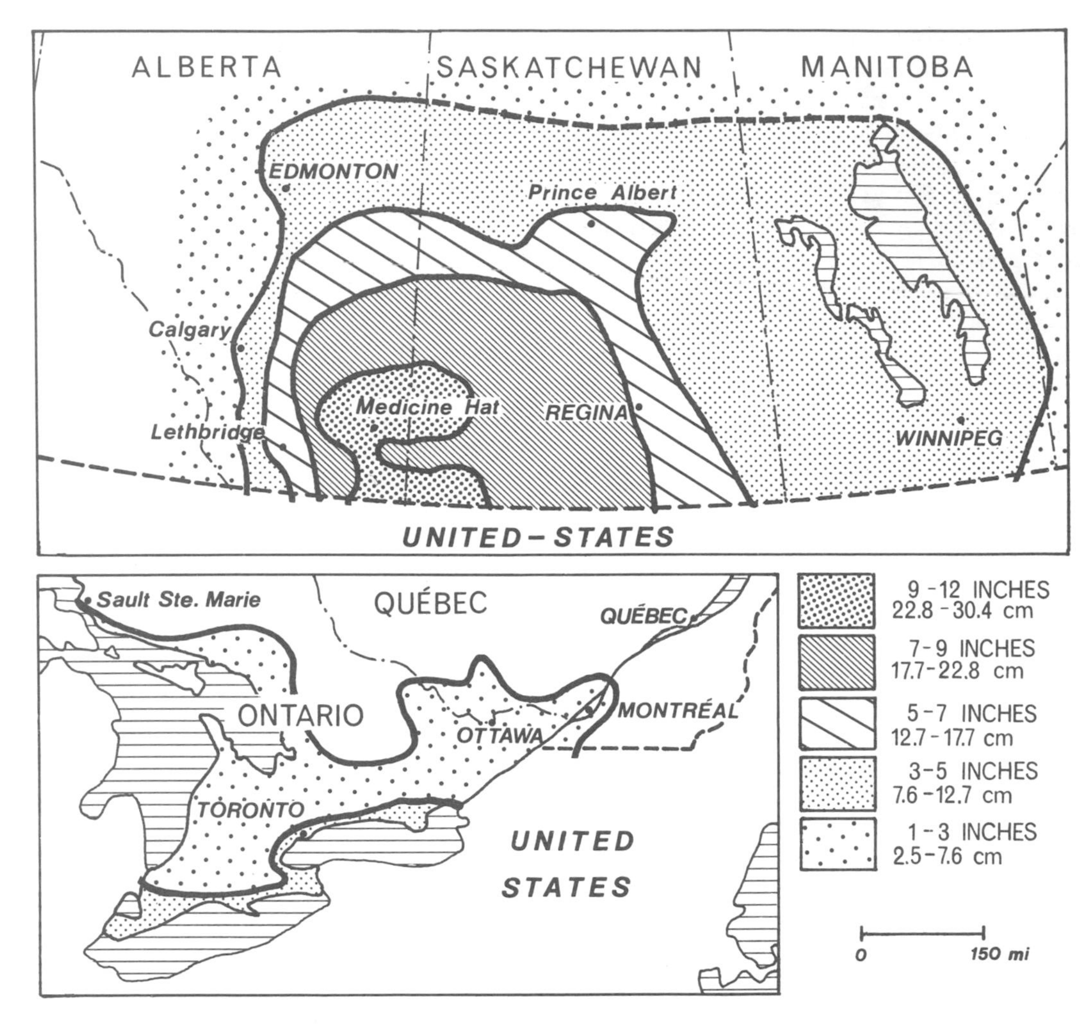

Fig. 3.1 Water deficit in the main agricultural regions. (Data from the Department of Agriculture, Ottawa.)

which is situated farther south than the Montreal Plain, has a deficit of about 2 inches of water during the growing season. These deficits cause problems for agriculture and grass, as the turf, without irrigation, takes on a burned appearance.

However, it is in western Canada that plants are really affected. The Prairies have experienced droughts in 1886, 1917–20 and 1929–38. *Drylanders* is the expressive title of a Canadian film about the Prairies. A. H. Laycock estimates that in the most critical areas precipitation should be 10 inches (25.4 centimetres) higher to assure optimal conditions for plant growth.[3] At Swift Current, Saskatchewan, the average annual deficit from 1921 to 1950 was 8.5 inches (21.5 centimetres). Agricultural production is

3. See A. H. Laycock, *Water Deficiency and Surplus Patterns in the Prairie Provinces*, Prairie Provinces Water Board Report, No. 13 (Regina, 1967).

very dependent on water; in the Prairies a variation of 3 inches of rain can make the difference between abundance and economic crisis. Irrigation thus becomes a necessity. The works on the St. Mary River in Alberta already irrigate 500,000 acres of land; the South Saskatchewan River Development Project (Gardiner Dam) will provide water to irrigate another 500,000 acres. About one-third of the deficit area in the West (15 million acres) could in fact be irrigated. Drought assists wind action and requires the use of special agricultural methods, such as fallowing and harrowing (instead of ploughing). Lack of water during the summer (fig. 3.1), like the cold of spring and fall, therefore restricts cattle breeding and the development of agriculture. The deficit is still greater in the deep valleys of the interior of British Columbia, although the area of dry land is much smaller than that on the Prairies. According to Thornthwaite's formula, Ashcroft experiences a deficit for seven months a year. Under such conditions, irrigation is not merely a useful device for increasing profits—it is a necessity. The orchards of the Okanagan, the vegetables and the lucerne for feed are artificially watered; these watering installations increase costs and encourage high output.

Curiously, because the rainfall maximum is in winter, the fertile plain of the lower Fraser requires summer irrigation.

The Far North, a near-desert in terms of rainfall, is also short of water. During the cold season the Arctic is a very dry land where even water for fire fighting can be dangerously lacking. To use the terminology of temperate regions, one might say that the reserves are very slow to recharge and that the "surplus" is always small. Summer is characterized by a moisture deficit, one reason for the poor plant cover. The same is true in the Mackenzie Basin: Aklavik, with a hot summer, lacks 6 inches (15.2 centimetres) of water from July to September.

The cold, therefore, is not the only thing affecting the environment; the summer warmth causes significant evaporation and the Prairies in particular are affected by a serious climatic handicap.

Hydroelectric Power

Electricity is a vital natural resource. This form of energy is used throughout much of the Canadian economy, for processing primary materials, manufacturing, lighting, heating and other domestic uses. The fact that the two major users of energy, Ontario and Quebec, have little coal or oil has facilitated the development of their electricity potential. Canada is among the leading countries of the world in terms of both installed capacity and per capita production. Quebec's research centre on hydroelectricity transmission is among the most advanced in the world. Despite its importance, however, electricity is not and never has been the principal source of energy; in the nineteenth century this was coal and since 1950, oil and gas have taken the leading roles, electricity ranking third, with 10 per cent of the total British Thermal Units consumed in Canada (fig. 3.2). In 1971, oil and gas met more than three-quarters of the nation's total energy requirement of 5,488 million B.T.U.s.[4] Nearly two-thirds of the installed power (hydro and thermal) is located in Quebec and Ontario; the percentage in B.C. is almost twice that in the Maritimes (table 3.1).

In Canada hydro is the more important form of electricity generation: in 1970 only 32 per cent of electric energy produced came from thermal plants. The regional variation is, however, considerable, the well-watered provinces of Quebec, Manitoba, Newfoundland and British Columbia (hydro proportion over 92 per cent) contrasting with Alberta, Saskatchewan and Nova Scotia (thermal production 62 to 80 per cent). In 1969, Quebec produced 44

---

4. See *Facts and Figures about Oil in Canada. A Quarter Century of Progress* (Imperial Oil Limited, 1972).

per cent of the hydro power in Canada, and Ontario, 48 per cent of the thermal power. The installed capacity is 40 million kilowatts and will exceed 55 million in 1974. Among the new gigantic projects, note the Manic group (opened in 1968) and Churchill Falls (1972); both are hydro installations and have a capacity of 5 million kilowatts. Ontario develops thermal and nuclear power at Bruce, Pickering, Lennox and Nanticoke. In the Middle North region of the Hudson Bay watershed, hydro plants have been developed on the Nelson River (Manitoba) and projects are under way on La Grande River in Quebec (part of the James Bay scheme).

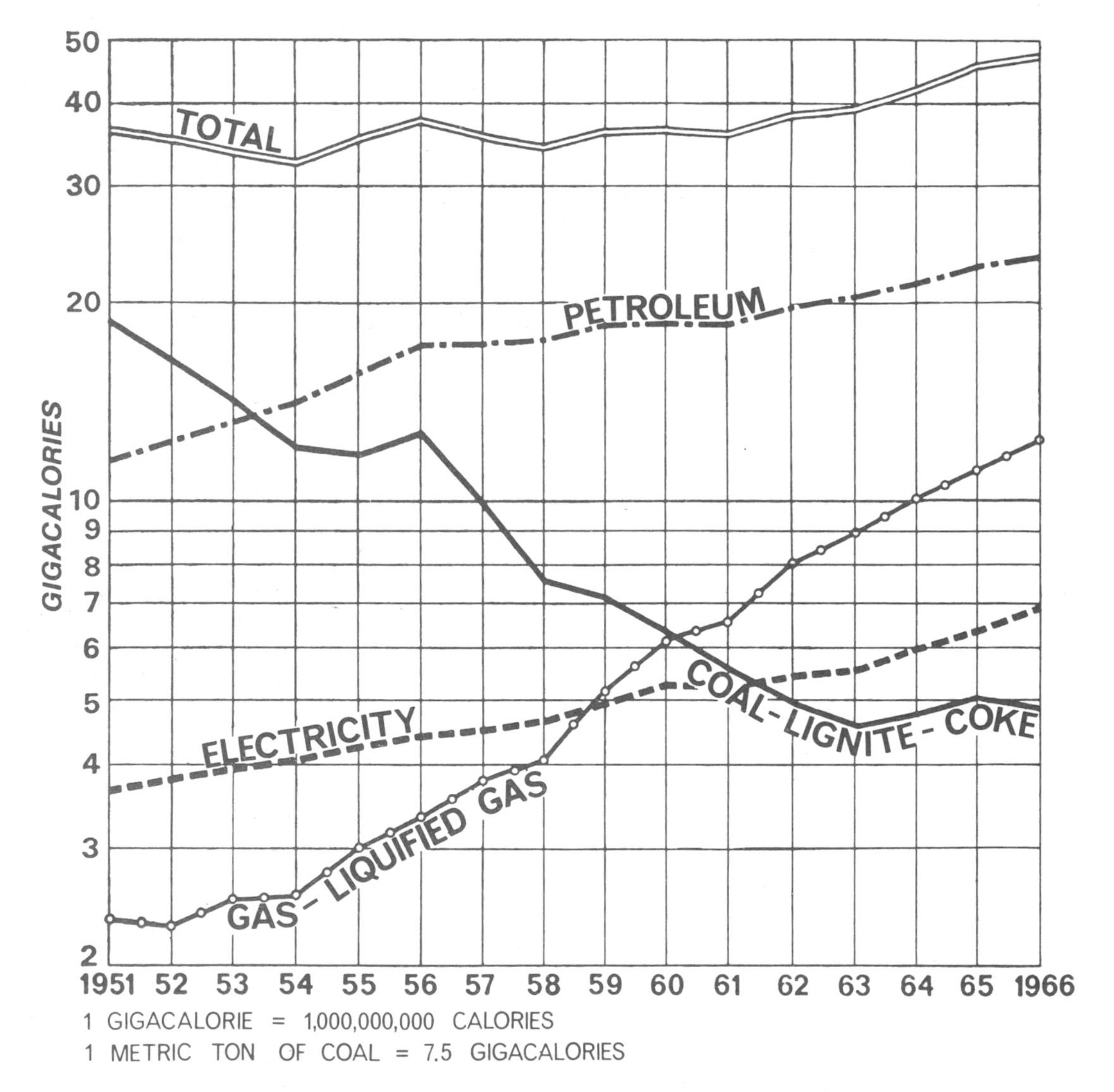

Fig. 3.2 Use of energy per capita. (Calculations by P. Bussières from data supplied by the National Energy Board.)

TABLE 3.1 Installed Hydro- and Thermal-Electric Generating Capacity

	1970 (%)
Atlantic Provinces	8.0
Quebec	33.3
Ontario	31.4
Prairie Provinces	13.5
British Columbia	13.6
Territories	0.2
Total	100.

SOURCE: [Canada], *Canada Year Book, 1971.*

Electricity plays an essential role in the economy and it is significant that its use and industrial activity have grown simultaneously. The three major provinces are those that produce the most electricity; their important industries, such as paper making and refining, could not operate without it. Juillard has accurately noted that the economies of the St. Maurice, Saguenay and British Columbia are directly dependent on hydro power,[5] which is available at low cost. Until the Churchill Falls development is complete (around 1975) Canada will continue to export very little electricity (3 per cent in 1963).

5. E. Juillard, *L'économie du Canada*, p. 86.

In view of the great needs of the U.S.A. and the immense resources of Canada, water could, in the near future, become a very profitable item of export or a major factor in economic development, or both. Because of this, Quebec and British Columbia, which are vast and relatively humid and therefore have enormous water resources, are in a very favourable situation in North America. A comprehensive policy for water in both regional and functional terms is therefore a vital requirement. This policy must also be intimately concerned with the protection of the environment, for the majority of rivers and lakes in southern Canada are polluted to some extent. Examples of this are Lakes Erie and Ontario and the St. Lawrence River upstream from the estuary.

Chapter 4

The Northern Domain

Although Canada is regarded as a northern country through its location and other characteristics, a real awareness of these characteristics is still generally lacking. Few authors are anxious to delineate the North, to fix its southern boundaries, to define regions within it or even to elaborate an appropriate terminology. It is essential, for example, that the North not be confused with the land inhabited by the indigenous peoples; many of the latter in fact live in southern Canada.

A Circumpolar Index

So as to set the northern character in a wider context, a set of ten criteria is used here to reflect both human and physical geography (table 4.1). Among them, natural factors (including heat) clearly dominate; if the principal characteristic of the North, the cold, is well represented, the index is not a climatic one. Psychological factors are not, however, used in the same way as direct measures of the northern condition, especially imponderables such as heroism, the spirit of adventure or of solitude, and real or imagined hazards. Nevertheless, the fact that four human elements are included enables a more accurate perception of the North than one based solely on criteria such as the tundra-taiga boundary, the vague boundary of continuous permafrost, or the 50°F (10°C) July isotherm.

TABLE 4.1 Nordicity Index: Criteria, Classes, Score

Criteria	*Classes*	*Score*
(1) Latitude	90 degrees	100
	80 degrees	77
	70 degrees	55
	60 degrees	33
	50 degrees	11
	45 degrees	0
(2) Summer heat	0 days above 42°F (5.6°C)	100
	40 days above 42°F (5.6°C)	80
	60 days above 42°F (5.6°C)	70
	80 days above 42°F (5.6°C)	60
	100 days above 42°F (5.6°C)	45
	120 days above 42°F (5.6°C)	30
	135 days above 42°F (5.6°C)	20
	More than 150 days above 42°F (5.6°C)	0
(3) Annual cold	More than 12,000 day-degrees F below 32°F (0°C)	100
	10,000 day-degrees F below 32°F (0°C)	85
	8,500 day-degrees F below 32°F (0°C)	75
	7,000 day-degrees F below 32°F (0°C)	65
	5,250 day-degrees F below 32°F (0°C)	45
	3,500 day-degrees F below 32°F (0°C)	30
	2,250 day-degrees F below 32°F (0°C)	15
	Less than 1,000 day-degrees F below 32°F (0°C)	0
(4) Types of ice:		
	Permafrost, continuous and 1,500 ft thick (457 m)	100
	Permafrost, discontinuous	60
(i) Permafrost	Frozen ground, for 9 months	50
	Frozen ground, for 4 months	20
	Frozen ground, for less than 1 month	0
	Permanent pack ice in the Arctic Ocean	100
or	Pack ice in the peri-Arctic (Baffin)	90
(ii) Floating	Pack ice for 6 months	40
Ice	Pack ice for 4 months	20
	Pack ice for less than 1 month	0

Criteria	Classes	Score
	Ice cap – 5,000 ft thick (1,524 m) and more	100
or	Ice sheet – above 1,000 ft (304 m)	60
(iii) Glaciers	Névé	20
	Snow cover for less than 11 months	0
(5) Total precipitation	Less than 4 in (100 mm)	100
	8 in (200 mm)	80
	12 in (300 mm)	60
	16 in (400 mm)	30
	More than 20 in (500 mm)	0
(6) Vegetation	Stony desert	100
	Thin tundra	80
	Thick, shrub tundra	60
	Open forest (subarctic, parkland)	40
	Continuous forest (coniferous)	0
(7) Accessibility other than by air (heavy transport)	No service	100
	Seasonal service:	
	(i) once a year	80
	(ii) for 2 months	60
	(iii) for 6 months, or 2 seasons	40
	Service throughout the year:	
	(i) 1 form only	20
	(ii) more than 1 form	0
(8) Air service	Nearest airfield – approx. 1,000 mi (1,600 km)	100
	Nearest airfield – approx. 300 mi (480 km)	80
	Nearest airfield – approx. 30 mi (48 km)	60
	Regular service, twice a month	40
	Regular service, twice a week	15
	More than 1 service a day	0
(9) Population:		
	None	100
(i) Number of inhabitants in a settlement	From 20 to 30	90
	Approx. 500	75
	Approx. 1,000	60
	Approx. 3,000	20
	More than 5,000	0
	Uninhabited	100
(ii) Population density of the region (100,000 sq mi or 256,000 km²)	0.01 persons per sq mi (0.004 per km²)	90
	1 person per sq mi (0.4 per km²)	70
	2.5 persons per sq mi (1 per km²)	50
	5 persons per sq mi (2 per km²)	25
	10 persons per sq mi (4 per km²)	0
(10) Degree of economic activity	No production and none foreseeable	100
	Prospecting, but not yet exploited	80
	Gathering, or extractive industries or crafts	50
	Large ore deposits, entrepot, or terminus	30
	Large secondary industries	15
	Interregional centre for multiple services	0

TABLE 4.2 Canadian Example of Local and Regional Nordicity, by the Criteria of VAPO

Criteria	*(1) Settlements*			*(2) Regions*	
	Schefferville (Quebec)	*Dawson (Yukon)*	*Resolute (Northwest Territories)*	*Centre of Hudson Bay*	*Interior of Keewatin*
Latitude	21	42	65	31	42
Summer heat	37	24	90	71	60
Annual cold	42	50	80	54	75
Ice:					
(1) Permafrost	60	60	92	—	75
(2) Floating ice	—	—	—	65	—
Precipitation	0	52	95	52	90
Vegetation	40	40	100	—	80
Accessibility other than by air	20	20	70	40	100
Air service	20	30	25	100	100
Population:					
(1) No. of persons	20	67	78	—	—
(2) Regional density	—	—	—	—	90
Economic activity	35	50	80	85	100
Nordicity total	295	435	775	622*	812

* Adjusted to a possible total of 1,000, for comparison with the landbased stations.

The index is designed to be representative of conditions throughout the year; it is not used to define a seasonal North. The old notions of a summer North and a winter North are replaced by the concept of an annual North. Such an approach makes it much easier to appreciate the specific characteristics of each northern area, and this in turn enables more detailed comparisons between different areas, as, for example, between the eastern and western sides of Hudson Bay. Further, the index can be used to define northern water bodies, which would be impossible if the sole criteria were those of vegetation or frozen ground. Eight of the ten criteria suggested can apply to land or sea areas; so as to ensure compatibility of results, these are adjusted to values per thousand, though for sea areas they are first calculated out of 800 (table 4.2).

Calculation of the index is quite straightforward and is based on the premise that each criterion quantifies the northern character to a different extent. Summation of each of the individual values gives the total "northernness" or *nordicity* for the place in question. Each of the criteria has a scale from 0 to 100; for example, at the Pole the "latitude" element reaches 100. Hence, with ten criteria, the maximum possible value is fixed at 1,000 VAPO (polar values or units),[1] as seen in table 4.3.

1. L.-E. Hamelin, "A Circumpolar Index", in W. C. Wonders, *Canada's Changing North*, pp. 5–15.

TABLE 4.3 Global Values of the Canadian North, in VAPO

Alert	878
Melville Island	865
Resolute	775
Barnes Ice Cap	732
Cambridge	690
Coral Harbour	662
Frobisher	584
Aklavik	511
Povungnituk	502
Chimo	459
Churchill	450
Dawson City	435
Churchill Falls	432 (before the dam)
Yellowknife	390
Long Range	350 (summit)
Fort Smith	343
La Ronge	340
Whitehorse	283
Gagnon	277
Moosonee	270
Northeast, Gulf of St. Lawrence	238
Grande Prairie	211
Prince Albert	178
Chibougamau	151
Sept-Iles	133

Southern Limits of the Canadian North

As is inevitable, boundaries are only subjective, since everyone defines his own North. For some authors, there is a Near-North: "The Laurentians are already the North," wrote Raoul Blanchard.[2] For others, the boundary has been the coniferous forest, crossed by pioneering south-north penetration axes which are not linked to one another. Some experts would make the southern limit of the North coincide with that of the tundra-taiga boundary. Other northern frontiers extend all the way from the U.S.A. border to the Arctic Ocean. The northern index can, however, be used to define more comprehensive limits.

We might define two southern limits of the Canadian North, a maximum and an average. The former consists of the *isonord* (a line joining points of equal northern units) of 0 VAPO which skirts the Erie-Huron interlake region. This is an extreme frontier which, given the present stage of Canadian development, is now of little significance. The second boundary corresponds to the isonord of 200 VAPO which runs from the northeast part of the Gulf of St. Lawrence through Chibougamau, Quebec, and then northeast of Lake Superior in Ontario through the Peace River district of Alberta to the coast on the boundary of Alaska and British Columbia. This fairly modest northern index (200 units) may be taken as the southern limit of

2. R. Blanchard, *Le Canada français* (Paris: Presses Universitaires, 1964), p. 9.

Plate 4.1 The aircraft and the highway have ended the need for sternwheelers on the Yukon River. A new historical museum and several of these vessels provide a permanent memory of a unique phase in Canadian history, the Klondike gold rush. Note the fluvioglacial terraces in the background.

the Canadian North. It includes about 75 per cent of the land area of Canada and hence, in relation to area, Canada is a very northern country. South of this northern Canada is *Base Canada*, an expression which we prefer to "useful Canada" or "southern Canada".

Latitudinal Zonation

The index of northern characteristics is not merely a means to determine the maximum northern area; it is much more useful in describing variation in nordicity from place to place. The index enables one to appreciate the increase of northern characteristics the farther away one moves from Base Canada. This increase is not uniform and it is not strictly latitudinal. The isolines are not evenly spaced; northwest Canada possesses the steepest gradient of nordicity as, within a few hundred miles, the forests of the lower Mackenzie give way to permanent polar pack. On the broad scale, variations in nordicity are considerable, since values of several of the criteria vary markedly from place to place. This leads to local deviations in the isonordic lines, of the order of 100 to 250 VAPO. Such spatial variations may create local enclaves. For example, areas of very high nordicity, like the mountainous southwest Yukon, may be located in a region of otherwise only moderate nordicity. However, these local deviations do not destroy a general latitudinal trend in which the principal isonords are those of 500 and 800. The individual regions are separated not by narrow boundaries but by transition zones several tens of miles in width.

The Middle North

This zone is found between the mean isolines of 200 and 500 VAPO; about 500 miles (800 kilometres) in width, it extends from the Labrador Sea to central Alaska. Climatically it possesses generally subarctic features. In the context of major ecumene types, the Middle North is an area of isolated pioneer ventures and sporadic settlement separated by vast unattractive regions. The main points along the southern boundary are the Strait of Belle Isle (Newfoundland), Sept-Iles (Quebec) and Waterways (Alberta); on the northern border are Chimo, Churchill and Inuvik. Included in the Middle North are Schefferville (on the 1927 Quebec-Labrador boundary), Moosonee (Ontario), Flin Flon (Saskatchewan-Manitoba), Fort Smith (Alberta–Northwest Territories), Yellowknife (Northwest Territories) and Whitehorse and Dawson (in the Yukon). Hudson Bay itself forms an immense transition between the Middle North and the Far North and can be considered as belonging to either. Future pipelines from the Arctic will run through the Middle North.

The Far North

This term, which has been used in descriptive literature for several decades, is used here in a specific sense: it represents a nordicity greater than that of the Middle North. The bounding isolines are those of 500 and 800 VAPO. This zone is less well defined than the preceding one, but it is generally accessible from the Atlantic through Hudson Strait. It is an ecumene which is half land, half sea, and is, in terms of climate, the "Low Arctic". It is a zone that has practically no land links to Base Canada. Frobisher, Cambridge and Resolute are the main settlements, where the administrative and military authority of the state comes in contact with recent gatherings of Eskimos who are experiencing rapid acculturation.

The Extreme North

The name indicates the still greater degree of nordicity: 865 units on Melville Island. This is a land of glaciers, of prolonged pack ice or of ground which is frozen to great depths. It is an empty land, beyond the present limits of the

TABLE 4.4 Population of the Canadian North (by zone), 1966

Region	*Number of Inhabitants*
Middle North	210,546
Far North	17,814
Extreme North	140 (est.)
Total	228,500

SOURCE: Calculated from census data.

main Eskimo territory. Climatically, this High Arctic is a true desert, apart from a few places like Eureka and Isachsen.

If the northern index is extended to other boreal countries, most of the Arctic Ocean and Greenland is in the Extreme North, whereas about two-thirds of Alaska and the U.S.S.R., as well as almost the entire nontemperate part of Scandinavia, belong to the Middle North. Objective comparisons can be made between opposite sides of the Arctic Ocean: Norilsk (U.S.S.R.) has a nordicity almost identical to that of Churchill; Novosibirsk and Narvik are similar to Edmonton. The Russian Franz-Josef Islands are scarcely more northern than is Resolute on Cornwallis Island. Compared to its neighbours, Canada tends to have higher northern values. Canada also has more land in the Extreme North and Far North than in the Middle North (fig. 4.1), which is a disadvantage. This is one of the main reasons for the small population of the Canadian North; there are only about 250,000 inhabitants, not counting the urban areas along the margin, such as Sept-Iles. Despite local anomalies, the values of the northern index can be regarded as inverse indices of ecumene.[3]

The Need for a Comprehensive Plan

There is no shortage of development plans for the North; the most recent was set out by the present federal minister of Indian and Northern Affairs.[4] But it is unfortunate that there is no overall northern plan that would affect all parts of the region, including the northern parts of the provinces, go beyond the analysis of a single natural resource (e.g., petroleum) and not be limited every time only to the period between elections. It must be recognized, however, that even southern Canada has little tradition of comprehensive planning that includes more than one region, more than one sector and extends over several years. This is certainly the greatest need of the North at present.

Conclusion

In latitude as in climate, Canada is truly a northern land; it is a kingdom of cold due to planetary factors and to the advective characteristics of the general circulation of the atmosphere. Canada, although fronting on three oceans, does not benefit greatly from the usual oceanic effects on temperature, mainly on the

3. L.-E. Hamelin, "A Zonal System of Allowances for Northern Workers: an Example of Applied Geography", translated by W. Barr, *The Muskox*, No. 10, 1972, pp. 5–20.

4. See J. Chrétien, *Northern Canada in the 70's*, Dept. of Indian Affairs and Northern Development Report and Introductory Remarks (Ottawa, 1972).

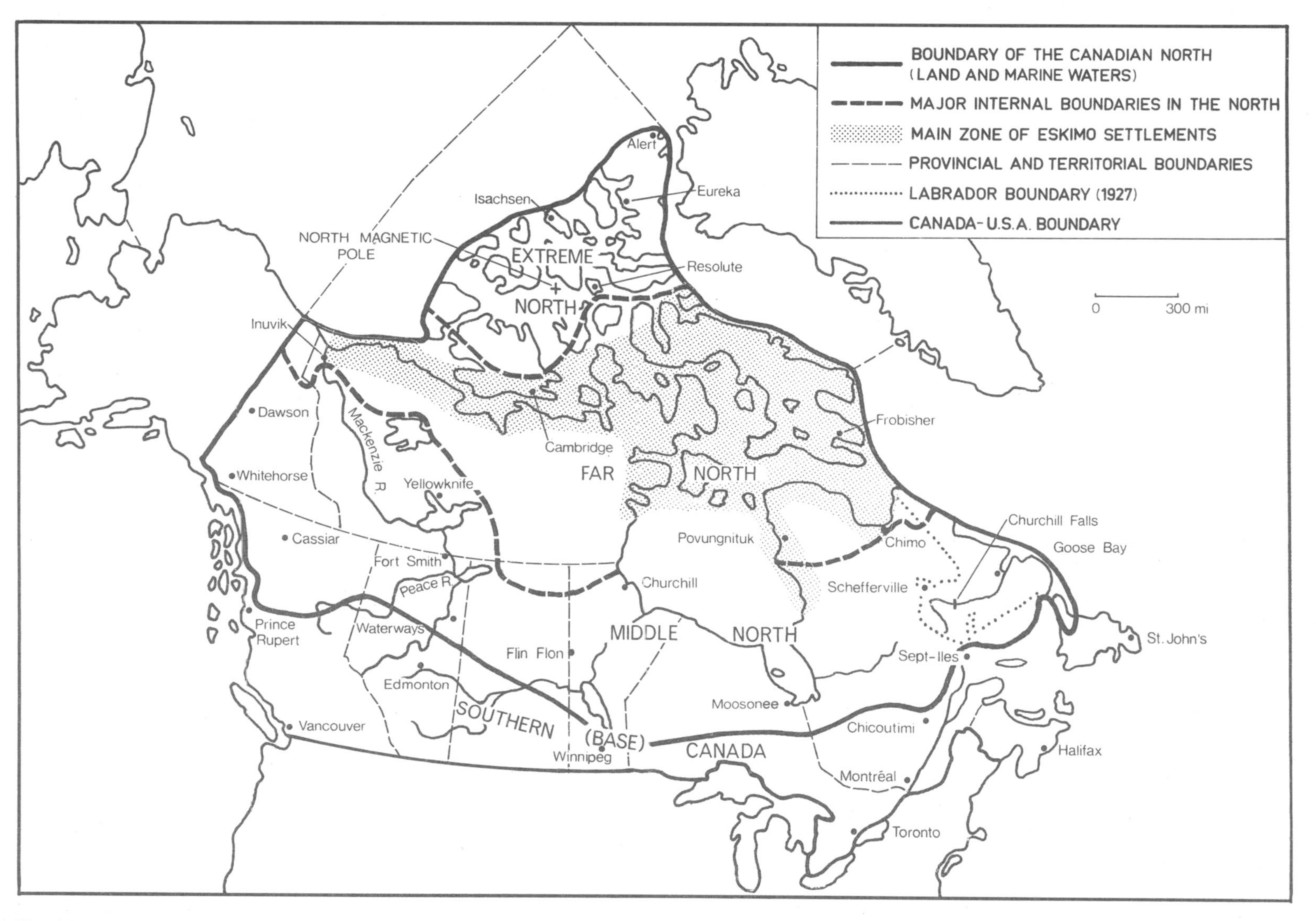

Fig. 4.1 Northern zones of Canada.

"good" Pacific side. Between Pacific and Atlantic and between the extra north-south barriers of Greenland and the Rockies, polar influences extend far to the south. This is particularly true in winter when freezing ice, joining together arctic islands and covering Hudson Bay, increases the continentality of Canada. Relief does nothing to protect Canada from the North, since the Canadian "Shield" is only a tectonic feature. At the surface the ground remains frozen for several months, reducing the period of plant growth over half the area of Canada; continuous and discontinuous permafrost represent both present and former cold. At the point of contact between cold air and frozen ground lies a snow cover, one of the most characteristic seasonal landscapes. The annual temperature cycle has three periods: growth (never more than four months), transition, freezing. The temperature range is everywhere so great that it must be referred to as changeable rather than as temperate. Six months apart, cold and heat both generate physiological stress.

Two Canadian regions have abundant moisture: the Western Cordillera, because of Pacific airstreams, and the Atlantic Provinces, due to invasions of subtropical air from the south. Except in the near-desert Arctic, the semi-arid areas of the Prairies and the interior of British Columbia, Canada is a fairly moist land with reasonable evapotranspiration values and a water balance that provides opportunities for energy production, forestry and agriculture. Despite the fairly well developed drainage pattern, several interfluve locations apparently are not drained either in winter or summer, a particular case of seasonal inland drainage.

Because of their unattractiveness, the hard and infertile lands of the North have significantly helped to confine human settlement to the southern periphery of the country or to the ocean coasts. The principal centres of population lie south of the Shield, in the Appalachian peninsulas and islands, or southwest of the Cordillera. The designation *southern Canada* describes the principal area of active Canada; as a consequence of its location, this Base Canada is likely to live under the economic shadow of the United States.

As one moves northward, annual receipts of solar energy diminish, negative thermal (freezing) index values increase, the opportunities of summer disappear, the snow cover (although becoming progressively thinner) lasts much longer, annual freeze-thaw cycles become generally fewer, the ground eventually becomes permanently frozen, forest gives place to woods, woods to rich tundra and this in turn to almost bare ground; the stream regime passes from a main low-water in summer to a long and distinctive low-water in winter.

Climatologically, Canada as a whole is a country very diversified and subject to seasonal changes. "No people on earth have a harder winter to complain about, latitude for latitude."[5] Pioneering ways of life still seem appropriate for it. In this infinite, varied and difficult land, however, there has developed a small southern ecumene where the majority of the 22 million inhabitants live in affluence.

5. F. K. Hare, "Canada at Large", in J. B. Bird, *The Natural Landscapes of Canada*, p. 1.

Part Two

Character and Regional Division of the Landscape

This part deals with those aspects of Canada that are most truly geographical: space or areal dimension. The country will be considered in terms of terrain conditions, spread of population, patterns of settlement, political structures and regional disparities. These considerations will show how and to what extent the country is basically organized.

Chapter 5

The Form of Canada

Structure

Until about 1950, it was common for writers to divide Canada into five vast regions, though there was less unanimity on what these should be called. They were termed geological regions, or physiographic, natural or mining regions. In fact, different criteria defined these regions. Bedrock outcrop defined the Canadian Shield; significant relief identified the Appalachians and especially the Rockies; low relief similarly defined the St. Lawrence and Hudson Bay lowlands, and vegetation was the distinguishing feature of the western Prairies. Recently the effective entry of the Far North into the geography of Canada has added a sixth, tectonic, element.

Recent attempts at synthesis by H. S. Bostock have enabled the creation of a preliminary map containing about one hundred structural units.[1] These so-called physiographic regions are based on both lithological and morphological criteria; they are not, however, morphological units in the sense in which the geographer would use the term.

The Canadian Shield forms the major unit, stretching from north to south almost without interruption over two-thirds of eastern Canada. This very heterogeneous complex of rocks, which has been affected by orogenies and erosion in varying degrees, covers 1,771,000 square miles (4,533,760 square kilometres). The Shield includes three major subdivisions, the first being the "Churchill Province", which

---

There is abundant literature dealing with the recent evolution of the natural landscapes. Among the many publications, the following are especially recommended:

(1) [Canada], *Short Papers on Quaternary Research in Canada*, collection of papers presented by the Department of Energy, Mines and Resources (Ottawa, 1969);

(2) C. Laverdière, ed., "Le Quaternaire du Québec. The Quaternary of Quebec", *Revue de Géographie de Montréal* 23 (1969): 225–392;

(3) S. Pawluk and J. Dumanki, eds., *Proceedings, Pedology and Quaternary Research* (Edmonton, 1969);

(4) V. K. Prest, "Quaternary Geology of Canada", in R. J. W. Douglas, *Geology and Economic Minerals of Canada* (Ottawa, 1971), pp. 676–764.

1. See H. S. Bostock, *A Provisional Physiographic Map of Canada.*

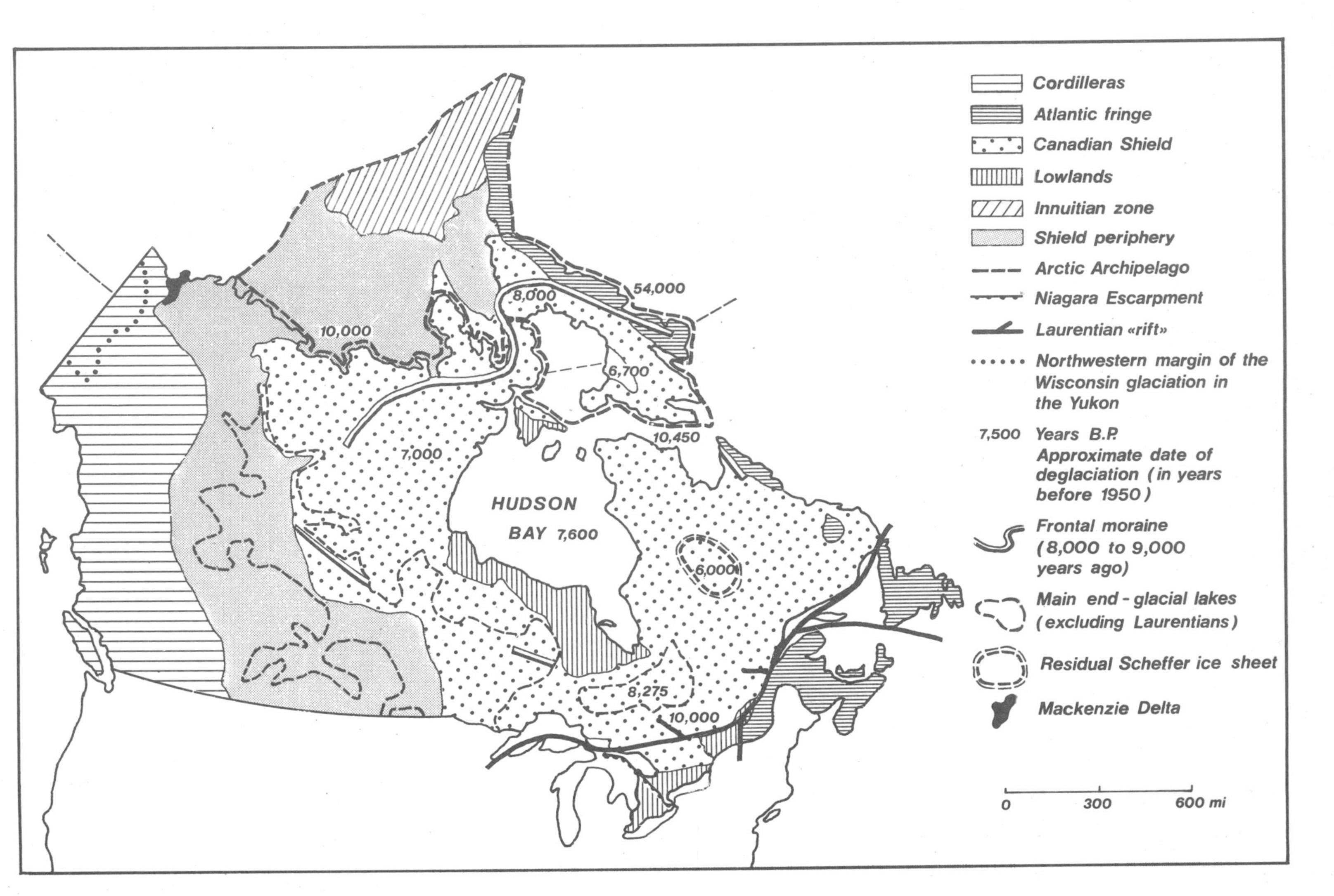

Fig. 5.1 Main physiographic characteristics of Canada.

surrounds the northern part of Hudson Bay and which is linked with the Labrador geosyncline. The latter is 525 miles (845 kilometres) in length, 60 miles (97 kilometres) wide and about 1.7 billion years old. The "Lake Superior Province" extends around the southern part of Hudson Bay and into a large part of Ungava; some portions are more than 3 billion years old. The young "Grenville Province" broadly coincides with the North Shore of the St. Lawrence; its abrupt edge, developed along fault or inflection lines and crossed by deep valleys, gives an appearance of modest mountains, to which the historian F. X. Garneau gave the name of *Laurentides* (or *Laurentians*).

Areas where the Precambrian bedrock is covered by later deposits form the second great structural region. Such areas, composed of deposits of Primary, Secondary or Tertiary age, are found around and within the Shield. An external arc is particularly well developed towards the west, but also extends towards the north and towards the southeast. Less symmetrically, this peripheral cover shows, in each of these three sections, a gradual increase in orogenic activity the farther one gets from the Shield. Thus the Cordillera is found beyond the Prairies, the Appalachians beyond the St. Lawrence Lowlands and the mountains of the Innuitian area beyond the low arctic plains. This similarity in pattern is, however, rather fortuitous; it is not a function of age, tectonics or the processes of erosion. The St. Lawrence Lowlands, so important in modern Canada, are composed of a cover of Quaternary deposits which in turn mask polygenetic surfaces. Only 42,000 square miles (107,520 square kilometres) in area, the Lowlands can be divided into two major areas, southern Quebec and southern Ontario, separated by the Frontenac axis which extends the Shield into the United States. Each of these two subregions is in turn divided by a significant feature, the Niagara escarpment in Ontario and the St. Lawrence River in Quebec.

The Shield is not, however, covered with deposits only along its outer edge. Pockets of such deposits remain here and there within it; for example, in the north (Foxe Basin), in the centre (Southampton Island), and at the southern end of the Hudson Bay depression.

Although the structure of Canada, therefore, resembles that of an onion, in terms of relief the country is saucer-shaped (fig. 5.1). The Hudson Bay depression occupies the centre; along the rim are found the Cordillera in the west, the Appalachians in the southeast, the Torngats in the east and the Innuitian mountains in the north. Between the Hudson Bay Lowlands and these high coastal rims lies a series of plateaux (the Shield) and intervening plains along the Windsor-Lévis axis and the southern Prairies. Rather as if this saucer were the product of a bad craftsman, the internal shape is not perfect. For example, the mountain barrier along the Pacific is more impressive than that along the Atlantic. Similarly, in western Ontario the physical symmetry is interrupted by the Canada-U.S.A. border, situated too far to the north. As well as being asymmetrical, this saucer is incomplete, a little like an archaeological find with some pieces missing. A raised edge is lacking along a large part of the southern rim and the intermediate section is riddled with depressions filled by huge lakes along the Shield edge, by the straits of the Arctic Archipelago and by the St. Lawrence estuary.

One of the major structural lines in the Laurentians, which extends from the western end of Lake Superior through Lake Nipissing and reaches the Atlantic through the two exits of the Gulf of St. Lawrence, is a rift valley like those in Africa or that of the Rhine.[2] Along this depression there occur *horst blocks* such as

2. P. J. Kumarapeli and V. A. Saul, "The St. Lawrence Valley System: A North American Equivalent of the East African Rift Valley System", *Canadian Journal of Earth Sciences* 3 (1966): 639–658.

Anticosti and the Adirondacks, or *tilted blocks* like western Newfoundland, Gaspé and Madawaska (Ontario), and *grabens* like the Saguenay, the Richelieu-Champlain depression, the central part of the Ottawa Valley and the lower Churchill River in Labrador. At one of the crossroads of this system are found the igneous intrusions of the Monteregian Hills.

Landform Evolution since the Last Glaciation

Before the first glaciation, Canada already possessed the main components of her present relief. In southern Saskatchewan, for example, geologists have discovered a complete preglacial drainage pattern. Erosion surfaces developed on the Shield, peaks and valleys in the Appalachians, the Cordillera and the arctic mountains, the level plains between the crystalline Shield and the folded rim—all these were already there and had been for a long time. In many places indeed, fragments of pre-Palaeozoic surfaces seem to be preserved. Pediplanation rather than peneplanation would perhaps be more probable in view of the tropical climates, both humid and dry, which succeeded each other in Secondary and Tertiary times.

As detailed knowledge of relief features associated with deglaciation gradually increases, it reveals the existence of a multitude of glacially dammed lakes with very irregular outlines. The majority are related to the shrinking of the large Hudson (Laurentide) Icecap and include

Plate 5.1 The glaciated landscape of the Shield north of Quebec City, Montreal and Ottawa provides excellent skiing from December to April. This giant slalom course is at Mont Tremblant, Que.

the Agassiz Lakes (8,000–11,400 years ago), 300 miles (482 kilometres) wide, which extended beyond southern Manitoba, leaving deposits over an area of 180,000 square miles (460,800 square kilometres). Similar depressions contained predecessors of Lakes Athabaska and Okanagan or the glacial lakes of the Barlow-Ojibway and the Lac des Cris in central Canada and the McConnel in the Mackenzie Basin. Many others could be mentioned, such as the former Lake Naskaupi, Lake McLean in northern Labrador and the Lake of Vermont, as well as many smaller features along the whole of the mountain edges or even around isolated hills such as the Monteregians. Even if most of the ice dams were temporary, they have nevertheless left behind lacustrine plains bounded by such characteristic features as deltas, like that of the Assiniboine. When these lakes emptied, they rapidly incised drainage channels, now long since abandoned.

Postglacial and Present-Day Landscape Evolution

After the disappearance of the ice, the depressed state of the land areas, combined with an eustatic rise in sea level, enabled Atlantic waters to cover the Canadian lowlands. Areas around Hudson Bay (such as the Tyrrell Sea), in the narrow valley of the St. Lawrence (such as the Champlain Sea and Laflamme Gulf in Lac St. Jean and the Saguenay) and the coasts of many of the arctic islands thus became submerged continental platforms. Engulfed by cold water several hundred feet deep, some of these lowlands were the scene of significant deposition represented today by a marine plain.

Subsequently, however, the continent recovered from the weight of the ice. It has been estimated that the recovery took place at an average rate of several feet per century (ignoring reversals and short periods of stability); at present the mean rate is only about 6 inches (15 centimetres). In the course of this emergence, the relative positions of seas and coasts were shifted and the former relief features caused by submergence were affected by processes such as coastal erosion and river and marine deposition. In many places, sandy deltas developed gradually outwards to cover marine clays. On Ellef Ringnes Island, rivers have extended their natural levees as the sea has gradually retreated.[3] Terraces, raised beaches and marine shells are the best evidence of the former shoreline. Polygenetic forms frequently occur, and immense parabolic dunes have developed on the beaches of the former hydrographic surfaces. Lastly, a youthful pattern of river valleys continues to develop on these recently exhumed, unconsolidated materials. In adjusting to the new base level, rivers have left longitudinal terraces along their axes. Whether this postglacial recovery has ended or not, some areas remain submerged in the eastern St. Lawrence Lowlands, in the Arctic Archipelago and in the Hudson Bay area. This marine incursion might justify the remark of the poet Gatien Lapointe: "The sea has taken my country by the hand."

Landslips have been one of the most active landscape evolution processes in recent times; they occur especially in areas with unconsolidated superficial deposits such as moraines, kame terraces and marine or lacustrine plains. When these lose their cohesion they become fluid; the innumerable valleys which extend over the surface of the St. Lawrence Lowlands are thus enlarged by such slips. In quite different circumstances, there occur in the Cordillera collapses of rocks, of snow and of covering deposits (such as at Frank in 1903 and Hope in 1965).

---

3. See D. St. Onge, *La géomorphologie de l'île Ellef Ringnes, Territoires-du-Nord-Ouest, Canada*, Étude géographique, No. 38 (Ottawa, 1965).

Valleys

Geographically, there are four types of valleys, which must be clearly distinguished because they are sometimes contained one within the other.

(1) The active "notch", generally narrow and little incised; its adjacent slopes mainly develop in relation to the stream which flows along its axis. This type of valley scarcely extends beyond the river banks.
(2) In the St. Lawrence Lowlands, enlargements several hundred yards wide represent the sites of former landslips, associated, among other things, with spring floods and meanders.
(3) The "Quaternary Valley", much too wide or too deep to have been eroded by present-day fluvial activity. Gorges some 300 feet deep, cut in hard rocks, began as overflow channels draining ice-dammed lakes.
(4) Lastly, on a different scale and begun much earlier still, the major depressions or troughs, such as those in the Laurentides and the Cordillera, which still occupy preglacial sites.

Valleys of the first type are often found within the second and third types, which in turn may be within the fourth, a vast infilled depression which, genetically, has nothing to do with the present drainage pattern. It may indeed be misleading to call such valleys by the names of the rivers that are found in them at present. Only the little valleys of the first type can be the product of fluvial action alone; the depressions of the second, third and fourth types are polygenetic. Settlement and communication patterns developed mainly in these larger units as well as along the low terraces of more recent valleys.

Lakes

Some day the geography of Canadian lakes will have to be written. The great lakes of the St. Lawrence and Mackenzie systems are obviously outstanding, but equally important is the total number of lake basins: there are probably a million in the Quebec-Labrador peninsula alone. A significant measure is the proportion of lake surface to the total area of a drainage basin; even on the plateaux of the central Shield the overall ratio is around 15 per cent, but it rises to 31 per cent in the 650 square miles (1,665 square kilometres) of the Opawica (Quebec), and it is more than 50 per cent around Liverpool Bay (Mackenzie) and in the basin of the Great Bear River. In the southern Arctic Archipelago, innumerable temporary lakes or sounds are often associated with glaciation and with *thermokarst*. Another useful index is that of length in relation to width; this index is about 14 for the lakes in the structural Labrador Trough; in the northern part of the Mackenzie such values are much less, even when limited to lakes with a definite orientation. Lastly, the insularity index expresses the proportion between the area of the islands and that of the inland waters in which they are found; the lakes with morainic or meltwater features contain many small islands but the insularity index rarely exceeds 30 per cent. The Lake of the Woods (Ontario) includes about 14,000 islands. In terms of lake features, one of the most interesting Canadian regions is the Prairies, where D. G. Frey has distinguished four major lake regions.[4]

Despite their frequency, lakes and rivers have done little to change postglacial relief; their effects are rarely more than a slight incision. In any case, fluvial erosion is not significant except during the spring run-off; it is mainly a seasonal phenomenon.[5] In Canada, therefore, the drainage pattern cannot be regarded as the key

4. See D. G. Frey, *Limnology in North America* (Madison: University of Wisconsin Press, 1963).
5. D. K. MacKay, "Characteristics of River Discharge and Runoff in Canada", *Geographical Bulletin* 8 (1966): 219–227.

Plate 5.2 A small part of the Mackenzie Delta, N.W.T.: a landscape that is as distinctively Canadian as the peaks of the Selkirk range (plate 16.2) or the Saskatchewan wheat-fields (plate 7.1). The maze of islands, secondary channels, ponds, bog and occasional patches of forest sustains a wide variety of wildlife, including abundant muskrat, beaver and waterfowl.

to landscape development, in the way that it can in areas that were not glaciated in the Quaternary and where fluvial processes have always been the major factor in landscape evolution.

Conclusion

What remains to be studied is far greater than what is known;* over the larger part of the country, we are still at the stage of a preliminary geomorphological reconnaissance. Some major problems remain practically untouched, such as the age of the extensive erosion surfaces, the origin of the continental shelf, the nature of the Arctic Archipelago, the vast Hudson Bay area, the existence of a double set of major lines of curvature along the edge and in the interior of the Shield, the pattern of Late-glacial events, the sequence of climates leading into the Pleistocene and the contribution of periglacial activity to Quaternary evolution.

In principle, the geographically trained geomorphologist tries to relate his work to man's activity. It is obvious that southern Canada reflects the influence of structure; the main settled area is found on either side of the Shield (southern Ontario and southern Quebec in the southeast, the Prairies in the southwest). Also, it lies mainly within the peripheral mountains (the Rockies and the Appalachians). Further, the relief features represented by the valuable natural gateways of the St. Lawrence and Hudson Bay enabled the continental interior to be opened up quickly, completely and by relatively small numbers of people. But relief is not the only factor to be considered when thinking of the habitability of a region. Of the numerous plains that are found throughout the area between the peripheral mountains and Hudson Bay, only those which are climatically suitable are areas of agriculture (scarcely more than 5 per cent of the total area of Canada). Further, the complex effects of relief change with the passage of time; the Canadian Shield, for a long time an excellent source of furs, discouraged settlement. However, in the twentieth century, technology and new needs have changed this apparently permanent role. Hydroelectricity developments, mines and forests in Labrador, Quebec, Ontario and Manitoba have led to the establishment of towns and villages. Hence the Shield, while staying morphologically the same, has changed geographically. As the historian H. A. Innis has written, the Shield, on which the first words in the economic history of Canada were written, could well have the last word.[6]

Man has changed his attitude towards the land he lives in. Rather than being just a framework, the land has become a friendly environment requiring the same controlled development that other resources do. Where there has been extensive destruction of the landscape, appropriate action must be taken. It is better perception of the environment that has led to the thorough private and governmental studies of the arctic environment being carried out before pipeline construction is authorized.[7]

---

* An excellent contribution to the knowledge of Canadian physiography has, however, been published recently – *The Natural Landscapes of Canada*, by J. B. Bird (Toronto: Wiley Publishers of Canada Limited, 1972).
See also M. Pardé, "Hydrologie du Saint-Laurent et de ses affluents", *Revue de Géographie de Montréal* 2 (1948): 35–83.

6. See H. A. Innis, *The Fur Trade in Canada.*
7. See [Canada], *Canada. Pipeline North. The Challenge of Arctic Oil and Gas*, Task Force on Northern Oil Development Report (Ottawa, 1972); and [Canada], *Land Use Information Series* (44 maps at 1:250,000 covering Mackenzie Valley and northern Yukon; further maps in preparation) (Ottawa: Department of Indian Affairs and Northern Development and Department of the Environment, 1972).

Chapter 6

Evolution of the Settlement Pattern

Discovery: Real and Official

The first group of discoveries includes those by native peoples and those by Scandinavians.

The migration routes and movements of the indigenous peoples, particularly those which occurred before the last glaciation, are not well understood. Very early migrations from Alaska could have taken place at the beginning of the Illinois (*Riss*) glaciation or during the first phases of the Wisconsin (*Würm*); in the latter case, such immigration would be dated about 65,000 years B.P. In the Yukon, excavations have determined some tools as being 27,000 years old.[1] A gigantic land bridge joined Asia to America during the last glacial maximum, allowing primitive man to enter Alaska and then to penetrate into Canada. This new migration towards southern Canada would have been able to make use of an interstadial at the western edge of the Laurentide glaciation. Much nearer our own time, after the last deglaciation, new waves of immigration reached America from Asia. There have therefore been several migrations moving from northwest to southeast.

Equally important, there have been movements in the opposite direction after the glacial periods, as native peoples have returned from the present area of the United States to Canada. The maize grown by the Iroquois at Hochelaga (Montreal) was a cultural characteristic derived from the south. Thus, over the course of several thousand years, Canada received waves of early immigrants coming both from Alaska and from the continental United States; the majority of these peoples intermixed with those already here. The supposed Alaskan origin of the people does not eliminate other hypotheses: some migrations might well have come from the east (Europe via Greenland) and even from the south (tropics via the United States). Whatever in fact happened, the descendants of these nomadic peoples were here to meet the European explorers and it is they who were the real discoverers and first inhabitants of Canada.

The other real discovery of Canada came as an extension of European exploration. The Sagas suggest that in 985 or 986 A.D. an Icelandic trader, Bjarni Herjolfsson, blown off course in the Atlantic, could have seen Nova Scotia and Newfoundland. According to some archaeologists, Leif Erikson (son of Eric the Red) sailed in 1002 to Newfoundland, which he called *Helluland* or 'the region of flat stones'. Then he went south to Halifax; seeing there forests, he called the region *Markland.* Still farther south, he called New England *Vinland* and wintered there. Experts do not agree, however, on the precise location of these regions. The Scandinavians established bases for exploration into the area which was to become Canada; for example, Knitson in 1363 sent an expedition from Rhode Island to Hudson Bay (*Ginnungagap*). Whether all this is true or not, the oldest-known map of America is the Vinland map of 1440. The Norse voyages, made in stages and dependent on the winds, linked Scandinavia, the British Isles, Iceland, and the southeast and later the west coast of Greenland, Baffin Island, Labrador, the Atlantic Provinces and New England. These first European visitors could even have penetrated into the peninsula of Quebec-Labrador. On the Labrador coast, Norsemen and Eskimos encountered each other with a resulting physical and cultural intermixing. Some historians suggest that before these

1. C. R. Harington, "Ice Age Mammals in Canada", *The Arctic Circular* XXII (1971): 66–89.

Vikings, there was a series of Irish voyages to America.

"Official" discoveries came much later. Since the colonial nations were rivals, there were as many claims to discovery as there were exploring nations, and indeed as many as there were expedition leaders. France claims Jacques Cartier as the discoverer of Canada in 1534; his voyages are echoed in Rabelais's writings of the navigations of Pantagruel and in *L'Heptaméron* by Marguerite de Navarre. Before him, however, Verrazano had sighted Nova Scotia in 1524. For their part, England and the merchants of Bristol base their claim on the voyage of the Italian John Cabot in 1497 and 1498. According to E. Brazan, it seems that, as far as Newfoundland is concerned, Cabot had been preceded by the Portuguese (perhaps by Diego de Teive in 1452, and certainly by João Fernandez and Pedro de Barselos around 1492).[2] Even the Gulf of St. Lawrence had been seen by the Portuguese and A. Fagundes entered it in 1521, thirteen years before Cartier. According to Marcel Trudel, the Portuguese established their first colony in America about 1520.[3] The first, and equally temporary, French outpost was established twenty years later. Official discovery of the Atlantic coast of Canada hence came about five centuries after the Scandinavian voyages. Cartier might more reasonably be regarded as the rediscoverer of the St. Lawrence Valley than as the discoverer of Canada.

Exploration of the land of Canada did not end there; it was a long adventure that lasted for four centuries and ended only with complete aerial photography of the country. Many people can claim, like the European explorers, to have discovered parts of Canada. The majority of these explorers remain practically unknown since they did not leave written accounts. History does, however, record the prolonged search for the Northwest Passage, the extensive and premature penetration of the continental interior, the exploration of the coasts of British Columbia, the waterways used by the fur companies, and the numerous British, Scandinavian, American and Canadian expeditions to the Arctic.

Settlement Frontiers

The early phases of settlement were separated by long intervening pauses. About a century passed between the formal discovery of the Atlantic coast by the Portuguese, British and French, and the creation of real colonies. Cabot reached Newfoundland at the end of the fifteenth century, but the official colonization, precarious though it was, only came at the beginning of the seventeenth century. Cartier was at the site of Quebec in 1535, but Champlain only founded it in 1608. Various reasons can be found to explain the slowness of colonization by France and England. The Canadian climate and scurvy were certainly major factors. Moreover, the colonies did not rapidly become a source of riches for the colonists. The chief reason, however, was that the European nations had other things to do. France was preoccupied with the Wars of Religion and France was more interested in European politics than in the New World, in any case. Colonial development also suffered as a result of the Reformation and from doubts about its own ultimate objectives: was it a mercantile enterprise designed to enrich middle-class investors, or was it to establish agricultural colonies with government support? In addition, France was undecided as to where to make the first experiments, in Acadia or inland on the St. Lawrence. Among other advantages, Acadian settlement would allow France to control the navigation at Canada's gateway and it would also be near the cod-fishing banks. The St. Lawrence River basin was

2. E. Brazan, *The Corte-Real Family and the New World* (Lisbon: Agencia-Geral do Ultramar, 1965).
3. See M. Trudel, *Histoire de la Nouvelle France, 1524–1603* (*Vol. I*).

more favourable for the fur trade, despite the Iroquois menace. It must be remembered too that America was not really desirable in itself; Europe was much more interested in finding, through America, the route to Asia. French and British colonization of Canada really only began in earnest in the eighteenth century. The early pace of settlement was slow indeed, proceeding in four main stages.

(1) *The Westward Movement and its Limitations*

All writers have naturally emphasized the gradual displacement of the frontier of settlement from east to west. This was true, however, only on the very broadest scale. It was not necessarily maintained among the smaller units: the seigneuries, townships, towns and their neighbourhoods. In 1961 nearly three-quarters of the Canadian population still lived in the east, so the spread of settlement has not been accompanied by a comparable shift of population. The median point of population in southern Canada is still located in Ontario. During the "age of Ontario" the St. Lawrence Lowlands prospered on wood, furs and naval construction. One area which has benefited fairly well from the development everywhere else is the Windsor-Lévis axis, where population, industries, ports and services have all shown major growth. The development of links between the different parts of Canada has helped mainly the economic growth of southern Ontario and Montreal. It is rather as if, during the development of the Canadian tree, some branches have been better supplied with sap than others. This differential growth has led to regional disparities so marked that they represent one of the greatest problems facing Canada today. We have to recognize the continuing dominance of the St. Lawrence axis. Despite the competition provided by the Hudson's Bay Company until 1821, Canada has been created and reconstructed again and again through this eastern river gateway. During the French regime, the limits of the diocese of Quebec were only those of the *portageurs*. Afterwards, even before the coming of the railway, the water routes designed to collect beaver fur enabled Montreal to drain even the Mackenzie Basin. The railway put western wheat into the winter ports of Halifax and Saint John. Industry and administrative structures have enhanced the dominance of Ontario and of southern Quebec. "Confederation was, in an economic sense, the result of the centripetal influences of the commercial organization of the St. Lawrence," wrote Albert Faucher.[4] Business attracts business; in terms of market, of economic security, and economies of scale, most new enterprises have found it desirable to locate in the best-organized and most densely populated part of Canada. The continental tail and the St. Lawrence Lowlands head have always been closely associated with each other.

(2) *The Waves of Settlement*

The pace of settlement, like the wind, does not always move at the same speed. We may distinguish between:

(i) the major waves, which were responsible for the opening up of areas the size of a province (such as Ontario after 1783), or of groups of provinces (such as the Prairies around 1890)

(ii) minor waves, such as those into Lac St. Jean (1838), the St. Lawrence Lowlands of Quebec in the middle of the nineteenth century, and the Peace River of Alberta at the beginning of the twentieth century

(iii) some short-lived waves, like those to the Cariboo in British Columbia around 1860 or to the Klondike (Yukon) in 1896

---

4. A. Faucher, *Histoire économique et unité canadienne*, p. 20.

Between these pioneering leaps forward the frontier of settlement remained practically stable. Since their period of massive immigration, for instance, even the relatively youthful Prairie Provinces have marked time. Seen across time, the advance of settlement did not in any sense follow a steady pattern.

(3) *A Halt to Agricultural Expansion*

In Canada, as in the United States, the area of ecumene has scarcely changed in half a century. The linear pattern of development into the more attractive parts of the Middle North is in no way comparable with the settlement either of Ontario during the nineteenth century or of the Prairies at the beginning of the twentieth. The fact that the settled ecumene no longer continues to expand to any significant extent affects the outlook of the individual. Canada can no longer be described as a new country in these terms and the end to expansion of settlement has led to the realization that economic prospects are not boundless.

(4) *The Changing Frontier*

All forms of the pioneer fringe were extremely variable in both space and time. Such variations might be seasonal (migration of the Indians, of lumbermen or of the inhabitants of coastal Labrador), or of long duration. Two principal

Plate 6.1 "Suburbia on the Shield" has become a commonplace of northern development. The unimaginative street pattern and the monotonous trailer housing at Churchill Falls, Labrador, can be justified by the brief existence of the settlement. Although the project employed several thousand persons in the construction phase, only a few hundred will live permanently at the site, and the mobile homes will be moved elsewhere. Note the landscape of boreal forest, developed in a subarctic climate.

reasons are significant in the quasi-permanent shifts of the settlement margin. If the frontier is agricultural, local conditions are so marginal that farming is barely possible and every other opportunity for work seems more attractive; this has occurred in the Ontario and Quebec Abitibi since the "golden age" of settlement during the depression. In the Prairies, new settlement and land abandonment are concurrent features. If the frontier is based on forestry, or even more on mining—like some of the incursions into the Subarctic—resource extraction is a matter of economics and especially of external markets so that the future of such pioneer fringes is decided very far from the frontiers themselves. A management decision, often taken abroad, may signify either the opening of new centres at a cost of hundreds of millions of dollars, or equally it may mean the abandonment of an entire town, as was the case in Uranium City. Then again, it may lead to stagnation, as in Abitibi. To appreciate such changes, the map of the pioneer fringe must be continually revised, especially where the fringe is not an agricultural one. On the margins of development, of travel and even of settlement, the presence of man is a phenomenon which can only be precarious and which is often temporary. As in Brazil, there is something fragile, unsubstantial, and vulnerable in the pioneer fringe of Canada.

Major Waves of Settlement and Canadian Unity

Settlement dynamics have not made Canadianization easy; instead of a growth in national characteristics, the successive settlement frontiers have tended to lead to ethnic diversity. Following the French-speaking settlers of the St. Lawrence came the British in the Maritimes, Ontario and in British Columbia up to the nineteenth century, and a variety of cultural groups in the Prairies in the twentieth century. Through these successive waves of different people, Canada itself took on new characteristics. This is not a country which grew by extensions of existing cultures into new areas. As each new wave of settlement occupied a different area, so a series of new Canadas came into existence, physical distance being matched by cultural distance.

Development of the Major Regions

In spatial terms the economic history of Canada can be divided into four major regions related to different periods of time.

(1) *Before 1780, the Age of the East.* Newfoundland, Acadia, New France, Hudson Bay and the settlements south of the Gulf of St. Lawrence constituted separate subregions, each of which found common and competitive opportunities in fishing and fur. Despite the creation of the Laurentian parishes, all these colonies more truly represented development and colonial prestige than successful settlement. The peninsular and insular pattern of settlement reflected both the coastal spread of settlement and its rather tenuous character in this first phase of colonization. Canada, even without taking the natives into consideration, was already biethnic.

(2) *From 1780 to 1867, the Age of Ontario.* The French defeat in 1763, and the American Revolution perhaps even more, forced Canada into a second phase. Colonization towards the southwest occupied the interlake region of Ontario. Outwards from what became Toronto, by a network of roads and then of canals and railways, a strong regional structure developed. Profitable trade ensued with the U.S.A., despite its republican character and its political ambitions towards Canada. The British character of the immigrant population encouraged the inclusion of Ontario in the British Empire. As much for external as for national reasons, a certain Pan-Canadian political feeling was born which benefited this province considerably. In

Ontario was also located the new capital, Ottawa.

(3) *From 1867 to 1915, the Age of the West.* In the few decades immediately after the American Civil War, a very different Canada came into existence west of Ontario. Western Canada was composed of two contiguous areas. In the first, the Canadianization of the lands previously held by the Hudson's Bay Company was accomplished in 1870. Through the railway, which arrived just before the grain farmer, the real colonization of the West was achieved.

Canada, already an exporter of wheat, became one of the granaries of the world, somewhat of a paradox for a cold country. This production shows clearly the importance of summer and the thermal gradient in Canada. If the preceding phase of immigration had been British, this one was multiracial.

Secondly, the other West—the Pacific or Far West—was already in existence as a maritime colony (Victoria) when the transcontinental railway reached the coast in 1886. Farther north, a railway to meet mining needs reached the Yukon around 1900. It was during this age of the West that the Canada of today became a political entity, as nine of the ten provinces entered Confederation.

(4) *The Age of the Towns.* Up to this point these major economic phases had corresponded to extensions of the territorial area. Now that the Pacific had been reached and southern Canada had taken shape, where could future economic growth be expressed in spatial terms? Continuation of the previous pattern would have pushed back the "last frontier", that of the North. In actual fact, the percentage of Canadian people living in the North has not changed significantly since 1900. During this fourth phase, there has also been concern less for the extension of the ecumene than for a better evaluation of it, with workers tending to move towards the most promising places. This period has unmistakably been the beginning of the age of the towns, for in 1911 they contained only 40 per cent of the total population. The majority of the urban population is now located within a Canada which had already been well defined much earlier, notably in southern Ontario and in southern Quebec. Until the middle of the twentieth century, manufacturing and port functions were the main bases of the urban population; since then, tertiary and quaternary activities have become increasingly important. Canada is so much in an urban phase that towns and their suburbs have really become the "frontier of Canada". This is even true in the Middle North, as illustrated in the small centres of primary resource activity. Such urban communities are spreading like fungi colonizing a dead tree. In a variety of ways they act as growth poles in regional development; together with the highways which provide the links between them they form the basic pattern of the human landscape.

The Present Ecumene

A discussion of ecumene should include reference to habitation, utilization, organization and the productivity of a territory.

Forms of Inhabited and Exploited Areas

"In certain cases, a broad correlation exists between the functional types of ecumene and their form. For example, an elongated form of settlement may be found along railway lines, whereas settlements associated with mineral exploitation sites tend to be of a dispersed oasis type. Beyond these, the ecumene takes four general forms.

"(1) The *bloc* form corresponds to an older type of settlement in which the physical environment enables expansion. Southern 'Alsama' (Alberta-Saskatchewan-Manitoba), with its vast area of cultivated land, furnishes the best Canadian example. On the regional level, the bloc type should have a diameter of approximately 50 miles, and if there are subecumene areas, they should occupy less than 10 per cent of the total area. The expression *bloc* does not imply that the population must be distributed uniformly within the ecumene, or that it should decrease evenly from the centre to the periphery; the term refers only to the form of spatial extension and it states nothing about the intensity of occupation.

"(2) The *linear* form is found in each of the chief functional types of the ecumene. Examples are found along the railway and highway systems of central Ontario, in the valleys of British Columbia, along the north coast of the lower St. Lawrence River, in the spread of suburban growth along major roads, and in some rural settlements in the Maritimes. By definition, the linear form must extend for about 50 miles along which interruptions are short. A special case of the linear ecumene is brought about by ring occupation around lakes, for example in the area north of Montreal; tourism is responsible for this subtype.

"(3) The *point* form is found throughout two-thirds of northern Canada. There are three subtypes.

(i) *Isolated point*. There, residential or exploitation centres are located hundreds of miles from one another and are not linked by road or railway. Eskimo settlements, polar radar bases, abandoned and old extraction sites (Rankin in the N.W.T.), and administrative centres (Inuvik) fall in this category. Isolation is an important characteristic because between these oases the land is almost completely devoid of man or at best it is a secondary ecumene.

(ii) *Linked point*. This subtype consists of exploitation or service towns situated along development roads; for example, Wabush and Schefferville, Churchill, and Whitehorse, where the arrangement of the settlements resembles knots in a rope. These northern settlements, like the isolated point settlements, can form over-organized villages, so much so that we have spoken of 'the instant creation of complete communities'.

(iii) *Temporary point*. This subtype is characteristic of a weakly developed ecumene and reflects either systematic nomadism or the partial disappearance of many marginal settlements.

"(4) The *dispersed* form of ecumene reflects, as does the preceding one, the discontinuity of the habitat. Whereas the point type corresponds to villages and small towns, the dispersed type refers to regions in which people are scattered over a few dozen square miles. In the case of southern Canada, the dispersed type is characterized at the local scale by frequent, multiform interruptions of the residential, exploitation, linking ecumenes. These regions can have two pioneer zones—an external margin and an internal margin. The subcategories here are numerous, and many sectors of southeast Canada and of the near North exhibit this type of bizarre-shaped and broken-up ecumene.

"The functional and morphological types furnish a systematic picture of the Canadian ecumene. In general, however, they do not correspond to any specific geographic locations. The characteristics of the habitation, exploitation and linking ecumenes are not associated only with the southern part of the country, nor are the interruptions of the habitat unique to the northern zones. Thus, these general types of ecumene should not be used as a basis for strict regionalization. On the contrary, the zonal arrangement of the Canadian ecumene has its own typology, and can be divided into four zones: southern ecumene (continuous and discontinuous), pioneer fringe, dispersed ecumene, and mainly uninhabited areas."[5]

---

5. L.-E. Hamelin, "Types of Canadian Ecumene", translated by R. M. Irving, ed., in *Readings in Canadian Geography* (Toronto: Holt, Rinehart and Winston, 1968), pp. 20–30. Originally appeared in *Proceedings, Royal Society of Canada.*

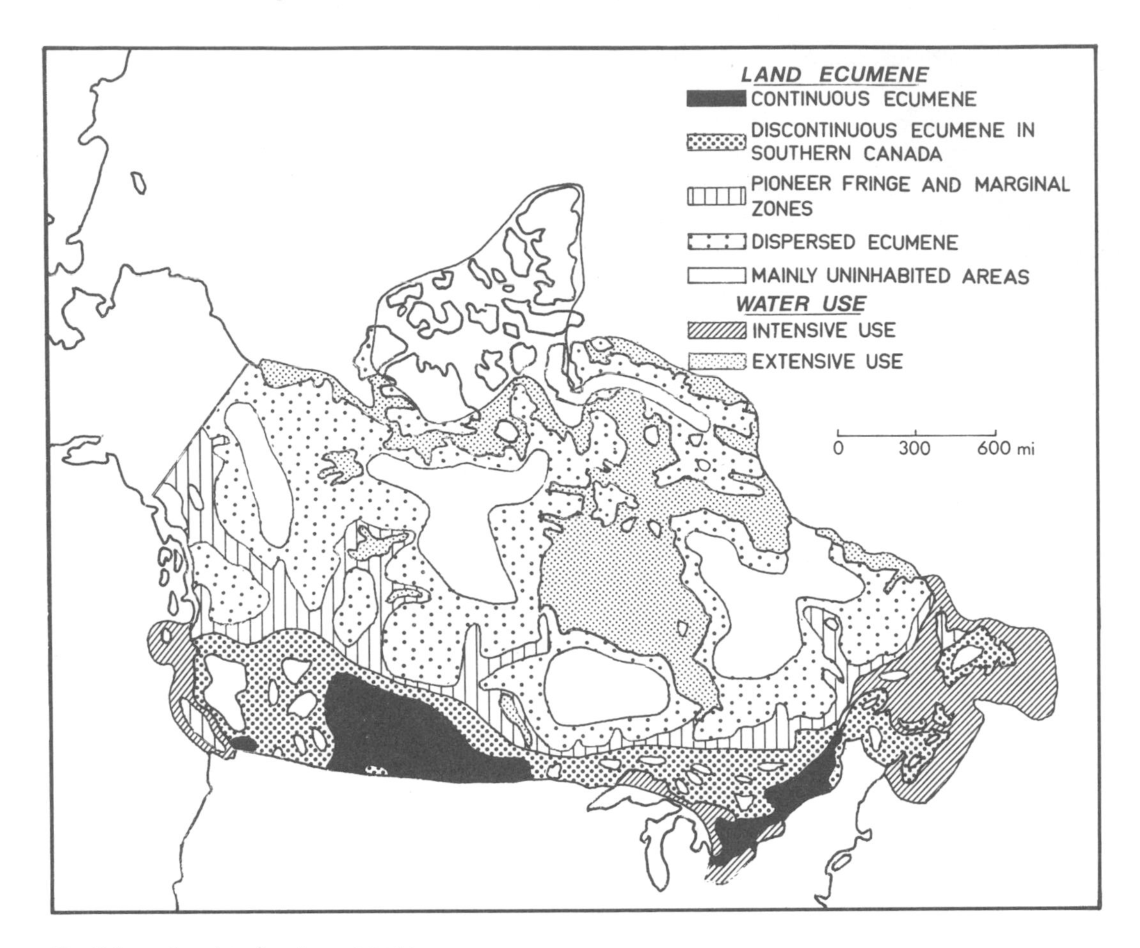

Fig. 6.1 Land and water ecumene.

Discontinuity

Even in southern Canada the ecumene is discontinuous (fig. 6.1). This is apparent first of all in terms of the major areas of settlement. For a long time there has been an unfortunate gap between the two main areas of Canada. An almost complete break in the line of settlement is the result of the Shield in northwest Ontario. This is the main cause of the division of the country into eastern and western Canada. In their turn, the Rockies divide western Canada into two areas of unequal size. The relationship of the settlement pattern to physical structure led H. Baulig to write: "The map drawn by nature has been faithfully followed."[6] Less a dominant factor, the Notre Dame area of the northern Appalachians breaks up the continuity of eastern Canada. As the historian A. R. M. Lower has emphasized, these breaks in the ecumene are "bad conductors" of the current of Canadian life.[7] Southern Canada is hence a collection of extending nuclei, separated by

6. H. Baulig, *Amérique septentrionale*, p. 273.
7. See A. R. M. Lower, *Colony to Nation. A History of Canada* (Toronto: Longmans, 1946).

intervening areas of thin or nonexistent settlement. From east to west, these settlement nodes are found along the shores of the Gulf of St. Lawrence (average population density of 45 persons to the square mile or 18 per square kilometre); in the Windsor-Lévis axis (80 to the square mile, 32 to the square kilometre); the southern Prairies (15 to the square mile, 6 to the square kilometre); and the lower Fraser (30 to the square mile, 12 to the square kilometre). Isolation tends to create as many sociological Canadas as there are centres of settlement. Spatial separation is reinforced by differing outlooks and economic development; as a result, communication between one area and another is difficult. Between these settled nodes, the emptiness persists and even southern Canada is still a collection of scattered pieces.

Throughout North America, the spatial dimension has affected the national pattern, influenced the form of settlement, and figured in the creation of attitudes of labour mobility, the squandering of land resources and the development of preferences by the early settlers.

Chapter 7

The Pattern of Rural Settlement

In the Old World, man as inhabitant or producer came before man as land surveyor; however, the opposite was generally the case in the New World. Almost everywhere in Canada, colonization began with a specific plan of land settlement, though this should not be confused with the official cadastral survey which only reached the provinces during and after the nineteenth century. So far as settlement is concerned, the township and the *rang* represent the ideas of the surveyors, not those of the settlers. Despite many instances where the lots are ill adapted to the actual facts of geography, such systematic models of land division have been one of the dominant structural elements of the rural landscape. Apart from their own intrinsic interest, the *rang* and the township demonstrate certain differences in the ethnic, chronological and functional characteristics of the agricultural areas.

For a detailed study of land surveys in Canada, see D. W. Thomson, *Men and Meridians*, Vols. 1, 2 and 3 (Ottawa, 1966, 1967 and 1969).

Seigneuries and *Rangs*

Two words which are intimately associated with traditional French Canada are the *seigneury*, an area which is subject to a specific set of administrative laws, and the *rang*, a pattern of settlement. In the words of Marcel Trudel: "The seigneurial system consists of the grant to entrepreneurs, termed *seigneurs*, of a substantial area of land for settlement, having established in advance and very precisely the rights and responsibilities involved, these being very carefully supervised by the state."[1]

In the St. Lawrence area, approximately 240 long and narrow seigneuries, the majority with one edge on the river itself, occupy a total area of about 8 million acres (3.2 million hectares).

Various assessments of the seigneurial system have, according to Cole Harris, exaggerated its importance; it did not have major economic,

1. M. Trudel, *The Seigneurial Regime* (Ottawa: Historical Society of Canada, 1956), p. 3.

social or geographical influences.[2] Other authors, however, believe that this framework safeguarded the French-Canadian identity since it was less attractive to English-speaking colonists than the freehold system used in the townships. The seigneurial system is only one element in a complex society and the main inconveniences of the seigneury (control on settlement and on population movement, lack of industrial development and, in the early phases, the absence of villages) must be evaluated within a much broader context.

The *rang* is *the* form of rural settlement in French Canada.[3] The seigneury is first of all divided into strips parallel to the river; the first strip is the first *rang*. The length of this section is 30 *arpents* (an *arpent* being 180 French feet, equal to 191 English feet or 58.2 metres). These *rangs* are, in their turn, rectangularly subdivided like the majority of the seigneuries themselves. The legal width of the unit was 1.5 *arpents* but the normal width is 3 *arpents*; the ratio of length to width is hence 10 to 1. The units of land subdivision are therefore rectangular, parallel and essentially similar to each other. From its origins as a cadastral arrangement, the *rang* has become a pattern of occupation. There are several types.

(1) The basic form is the *rang simple*, containing only one line of houses (usually built on the same side of the road).

(2) Planned settlement favoured the *rang double*, with the service road passing between two lines of lots; the overall population density in the parish is the same as in the *rang simple*, but the actual settlement sites are more concentrated and the road which services them is better used.

(3) The *rang-route* occurs on the margins of the Laurentian region, where, due to historical and topographical factors, the rural settlement pattern becomes a complex series of little *rangs* along which are found marginal farms or sections of roads without houses.

(4) The *rang-rue* of the nineteenth century is where the *rang* of the village, church, mill or railway was extended. The farm lots were divided so as to allow for a simple pattern of population concentration. Because of a natural tendency for extension along the street rather than on parallel streets, the villages became much longer than they were broad. Since World War II, such extensions have been especially significant around urban areas. Near Quebec, for example, the *rang de St. Jacques* (1696) became the rue de Neuvialle (1964); in Montreal, Metropolitan Boulevard is an extension of Côte de Liesse. Gross density of population rises to exceed hundreds per square mile. Despite urbanization, the old *rang* structure is still very apparent, even in the heart of the Quebec cities.

The value of the *rang* can only be assessed on the basis of very detailed analysis. Was it and is it a good system? Without any doubt, its regularity and its universal use were taken too far; it would have been better if soil quality and topography had been taken into account. Some *rangs* climb up the sides of the Laurentides, other *rangs* cross lakes, peat bogs and deep gullies. Such absurdities scarcely indicate a close relationship between man and the natural environment. It is also unfortunate that it has taken so long to reach the stage of the *rang double* and for villages to develop.

The *rang* did not encourage increases in the resident population; since the lots were usually indivisible, the farmer chose one of his sons to succeed him while the others were forced to go elsewhere. The *rang* therefore tended to encourage emigration, in the nineteenth century mainly

2. See Cole Harris, *The Seigneurial System in Early Canada. A Geographical Study*.
3. See P. Deffontaines, "Le rang: type de peuplement rural du Canada français", *Cahiers de Géographie*, No. 5 (Quebec, 1953).

to the U.S.A. and since then to the urban areas of the St. Lawrence Lowlands.

The extreme length of the lots and the fact that the home is at the extremity of the holding meant that an enormous amount of time was consumed in moving people and animals and in the construction of fences.

The farms, during the period of colonization, were at once too large for the scarce resources of the peasant family and too small to enable the rapid occupation of a huge colonial territory. Today, though there are tendencies both towards a reduction in lot size with more intensive cultivation of lots and also towards an increase in size, the latter is dominant. Regrouping of adjacent lots reduces the length-to-width ratio; alternatively, a fragmented farm operation is created. Some *rangs* are returning to forest, a time-consuming but renewable form of land use. In many settlements, the number of abandoned farms now exceeds the number of lots which are still occupied. In the whole of Abitibi in July 1967, only 19 per cent of the settled lots were still in a state of reasonable agriculture. More and more, those who live on the *rangs* earn an increasing proportion of their incomes off the farms. This "population pressure" is not solely a function of soil fertility, since the population density on the poor *rangs* of Bonaventure is higher than that on the fertile *rangs* of the Montreal Plain.[4] While the *rangs* close to the cities and towns experience an increase in urbanized population, *rangs* in more remote areas are losing their agricultural population, including both the excess and the essential components. Though *rangs* still remain on which the dominant function is agricultural (up to 80 per cent in the best parts of the Montreal Plain; 66 per cent for the good *rang* of "Pied-de-la-Côte" in St. Viateur–St. Barthélemy–Maskinongé), the *rang* is now in a period of major change. It is no longer the framework of protective conservatism but an instrument of evolution.

Townships

A *township* can be either of two things. First, it has an administrative or political connotation; in England the township represented the simplest form of administration. This was also the case in the early history of the U.S.A., where a township could support a church, a school, a military unit and a meeting hall. Such an area formed a local administrative unit and a certain sense of community developed within its limits. The second sense of the word is cadastral. The American type developed as a response to a desire to adopt a method of land division appropriate for the systematic development of the West. The model is in the Land Ordinance of 20 May 1785 and Thomas Jefferson was among those who developed it. A little later, the Congressional Township became a definite system with specific characteristics. The basic township, 6 miles (9.6 kilometres) square, was divided into thirty-six sections, each of 1 square mile (2.5 kilometres). These sections were further subdivided into four lots (fig. 7.1), the area of a lot being considered sufficient for the agricultural needs of the immigrants.

The Eastern and Abitibi Townships

The American Revolution did not in itself have an immediate and decisive effect on the settlement of the St. Lawrence Lowlands, since Governor Haldimand forbade immigrants to settle land along the boundary between Quebec and the new United States. Soon after, however, the land market in Quebec experienced one of its greatest booms. A systematic survey of the lands situated between the southern limits of the seigneuries and the American frontier—the region which became the Eastern Townships—

4. See L. Trotier, *Répartition de la population du Québec en 1961* (Quebec: Conseil d'Orientation économique, 1966).

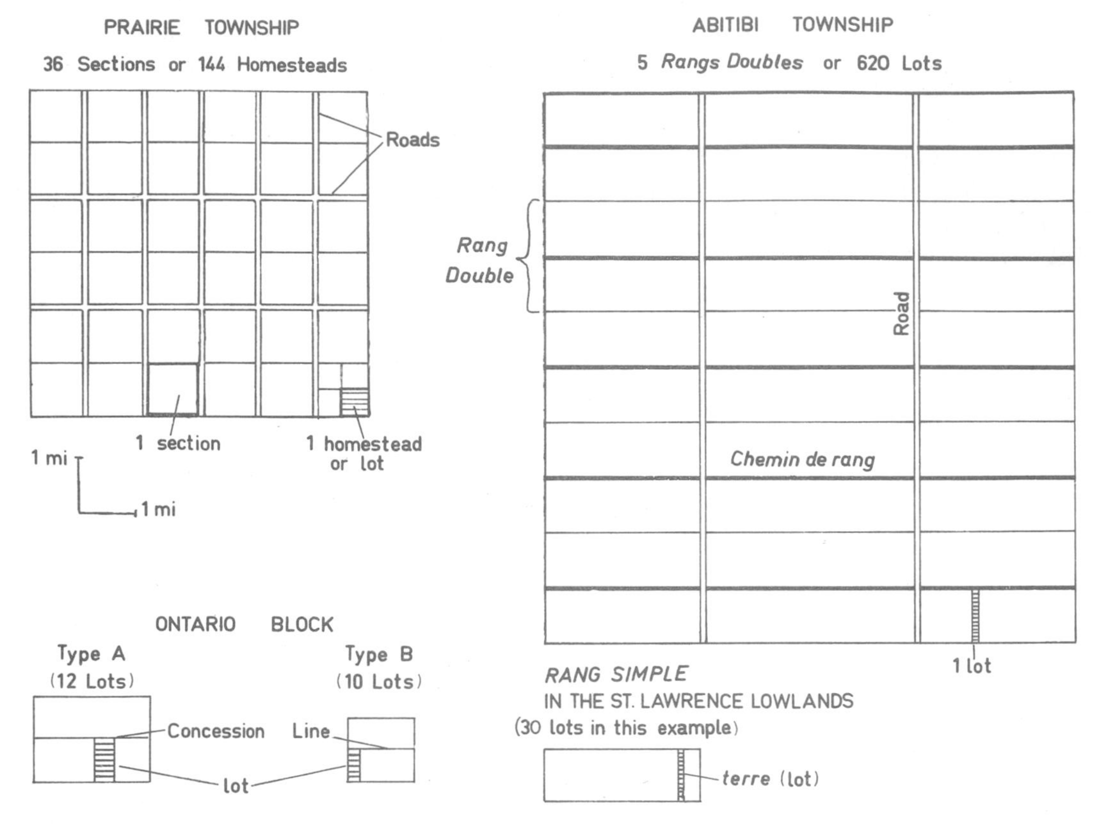

Fig. 7.1 Main settlement patterns.

was begun in 1792. Similar work went on in several places on the St. Lawrence Plain. By the time of an 1801 report, 188 more-or-less surveyed townships were mentioned, whose total area was about 20,000 square miles (50,000 square kilometres) or 12.8 million acres. In general these townships were 10 miles (16 kilometres) square or, on a waterway, 9 by 12 miles (14.4 by 19.2 kilometres). One of the methods of subdivision assumed ten *concessions*, each containing twenty-eight lots of 200 acres (80.9 hectares). The township and the homestead in the East are larger than those created by the U.S. Congress. The stage after surveying—that of settlement—was neither as simple nor as rapid as had been expected; in the Eastern Townships, real colonization only began a third of a century after the British victory in 1759.

The vast majority of the township heads were English-speaking: Republicans, United Empire Loyalists, or Britons with varying periods of residence in Canada. Those who could already be called *Canadiens* were little attracted by the townships, either in the land market or in settlement. Extension of the traditional parish framework beyond the old seigneuries was not yet definitely assured, causing French-speaking Catholics to stay aloof from these developments. It was only much later that the French

Canadians recolonized the Eastern Townships, on the "heels of the English".[5]

Other types of township have developed in Quebec, and one of them is widely distributed in Abitibi. Wholly French in character, there is a block of about 250 townships (*cantons*), each 10 miles (16 kilometres) square. Only a part of this block (45 *cantons* in 1961) has in fact been settled. The *cantons* were subdivided internally in a manner designed to preserve characteristics of the traditional landscape of French Canada. For example, the tendency of the townships bordering the St. Lawrence Lowlands to have lots oriented north-south (or northwest-southeast) is frequently repeated in Abitibi. The Abitibian townships in their purest form are subdivided into ten parallel bands aligned east to west; these are the *rangs* of the future. The length-to-width index of the lots is about 6 to 1; that is to say, about halfway between the (original) lots in the Eastern Townships and those of the seigneuries. There are about twice as many lots in an Abitibi *canton* as in an Eastern Township; the lot area is hence much smaller, and has proved to be too small to sustain highly profitable agriculture.

The Ontario Block

Ontario has also experienced a variety of township forms. In 1783, in response to the needs of the Loyalists, five units were surveyed in the Kingston area. During the first periods of settlement in Ontario, the township was a square area with sides varying between 6 and 10 miles (9.6 to 16 kilometres) in length. This survey was not just a cadastral exercise; where the unit included a parish, it exercised an administrative function.

In 1829, the block (part of a township) was fixed at 2,400 acres (971 hectares). This rectangle, surrounded by access roads, was about 2.25 miles (3.6 kilometres) long and 1.66 miles (2.6 kilometres) wide. The concession lines having been fixed beforehand at 66.67 chains (1,338.6 metres), each lot was therefore of that length. Since their breadth was only 30 chains (603 metres), there were six lots to a concession and twelve lots of 200 acres (80.9 hectares) in a block. These lots were therefore not as long and narrow as those of the Eastern Townships. Another type of block was composed of ten lots.

The Township of the Prairie Provinces

In 1869, federal Minister of Public Works McDougall asked Surveyor Dennis of Ontario to suggest a system of land survey appropriate for the Northwest Territories. The townships would be numbered northwards from the U.S. border on the 49th parallel; the ranges would be numbered east to west from a prime meridian situated 10 miles (16 kilometres) west of Pembina called the meridian of Winnipeg. The township would be square and the farm lots would be 210 acres (84.9 hectares) in size. Despite the advantages of the federal proposal, Lieutenant-Governor Archibald of Manitoba chose instead the dimensions of the lots in the American township: sides 6 miles (9.6 kilometres) in length, divided into thirty-six sections of 640 acres or 1 square mile (2.59 square kilometres). The section was subdivided in its turn into four homesteads of 160 acres (64.7 hectares) or into sixteen local subdivisions of 40 acres (16.2 hectares). The following example shows how a local subdivision was identified: 5 (the subdivision)–6 (section)–54 (township)–15 (range)–W 4 (meridian).

This system, which already included some differences from the American township, was modified later so as to find a better solution to the problem of the convergence of the meridians. This method of land division was very widely used in the Prairies and rows of about

5. See R. Blanchard, *Le Centre du Canada français*, pp. 339–360.

100 townships quickly extended from the American frontier all the way to the 60th parallel, the northern limit of the western provinces. The township pattern covered 400,000 square miles (1,036,000 square kilometres), or 20 per cent of the land surface of the ten provinces. It goes without saying, however, that the surveyed townships have not all been occupied.

Other Types

Townships and *rangs* have not been the only cadastral forms to leave their imprint on the Canadian landscape. Apart from the characteristic subdivisions of urban areas, mining concessions, Indian reserves and "squatter" settlements, there is the rare convergent triangle around Charlesbourg (Quebec).

A number of other types of townships deserve mention. In southeastern Canada, where colonization has been slow—or, in the case of Newfoundland, almost forbidden—cadastral division has also been slow, of only local importance and strongly influenced by American example. In the Maritimes, the unrewarding land and only limited areas of good soil, added to the intermittent character of the colonization and economic difficulties, has meant that the

Plate 7.1 Wilcox, south of Regina, Sask.; the familiar prairie township and section pattern. Prairie residents frequently try to convince other Canadians that "the prairies are not all flat". Although this is true, southern Saskatchewan is very close to the popular image. Trees are found only near settlements, usually planted as windbreaks. The number of "elevator towns" or service centres like Wilcox in the province declined from 909 in 1941 to 779 in 1961 and may fall to about 600 by 1981.

Plate 7.2 The unusual and distinctive radial cadastral system of Charlesbourg, as it was in 1950 before the spread of urbanization converted it into a suburb of Quebec City.

establishment of townships was not a characteristic of settlement as it was in southern Ontario and even more in the Prairie Provinces.

In British Columbia, the first important land survey dates only from 1858 and, because of the mountainous character of the province, only the extreme northeast and the southwest had surveys which were reasonably comprehensive. In general, a township pattern modified from that of the Prairies was used; on Lulu Island, for example, the townships are only 3 miles square.

Comparison of Settlement Patterns

Townships and *rangs* are primarily different in land division. We may take as a comparison the simple *rang* and the Ontario block, together with some sections of a prairie township. The *rang* is subdivided into *terres* and the section into homesteads. Originally, township lots were larger than those in the *rangs*: about 67 acres (27.1 hectares) for the lots in the Laurentian *rang*, 100 acres (40.5 hectares) in the Abitibi *rang*, and the same in the township of southwest Ontario. The prairie township lot was 160 acres (64.7 hectares) and in the old Eastern Township it was 200 acres (80.9 hectares). This original cadastral division is still reflected in areas of farms included in census returns. The class interval between 80 and 136 hectares is the dominant one in the Prairies but it is the interval from 24 to 44 hectares which dominates in Quebec. The lots no longer have the same length-to-width ratio. Ratios between 20 and 6 occur in the *rang* system, with a mode of 10; however, values may reach as high as 100, as in the old settlement of Beauport, Quebec. For a township, the index is from 2.2 to 2.5 in southern Ontario, and was originally 1 in the Prairies. The lot in a township is thus more compact than in a *rang*.

The land division pattern has led to differences in occupation. In the *rang* system, the farms are located at one end of the lot. In the township, they are a little farther from the service road; on the Prairies, houses are found not only along the east-west roads, but also along those from north to south. Hence, whereas the township almost everywhere led to a dispersed settlement pattern, the *rang* produced a linear settlement. The gross density of agricultural population is three times higher in the Laurentian *rang* than in the prairie township, a result of the form of the settlement as well as of the period of colonization.

In eastern Canada, forest clearance was rarely as complete as it was in certain sections of the Montreal Plain and in the Ontario peninsula. Soil character, the small time that has elapsed since the original settlement, the extensive character of the agriculture and the need for individual farm woodlots have all tended to limit destruction.

Even the geography of individual fields reflects the form of the cadastral survey (i.e., the straightness of a *terre* in the *rang*), the method of cultivation and the functions of the different units. There is a difference between vast and small fields. In the former, their use for cereals, cash crops, or for stock grazing determines the size of the units. On the average, the fields of the St. Lawrence Valley and those of Lulu Island (British Columbia) are about 10 acres (4 hectares). They are rather smaller in Abitibi, but in the Champion area of the Prairies agricultural units are about 70 acres (20.3 hectares). The small fields, usually situated near the farmstead, are used for poultry farming, cattle rearing and a variety of small cash crops such as strawberries. The average number of fields per farm in the St. Lawrence Lowlands is about ten; specific, amusing and sometimes spicy names are given to each of these units. As for the field shape, it is more irregular in the Maritimes than in the Laurentian *rangs* and the prairie townships. Problems of fencing arise from the need to mark the limits and to retain farm animals. Even though the normal fence material has changed from wood to barbed wire, the maintenance of fences is a difficult task, due in part to the periodic freezing of the ground and to the mass movement of the fine glacial marine or lacustrine deposits which are typical of many fields in southern Canada.

Chapter 8

The Creation of the Political Units

Land Acquisition

Even though nonindigenous Canada developed entirely within the period of written history, we still do not know all the methods by which the land area was occupied by Canadian society and economy. The most common method of land acquisition was certainly that of discovery, followed by declarations of possession. This happened not only during the French and British colonial periods, but even in the twentieth century with Stefansson, who made some discoveries, and with A. P. Low and E. Bernier, who took possession of the North. The majority of the other methods by which the land was acquired are specific to certain areas of Canada. The occupation or the development of separate colonies (Newfoundland, Acadia, New France, Upper Canada, the Selkirk Grant in Manitoba, and Vancouver Island) have been the most well-founded claims. The transfer of native lands to the white man was not a deed of shining virtue; nevertheless, a series of treaties mainly in the nineteenth century enabled the acquisition of part of the Indian lands of Ontario and the Prairie Provinces. Through these treaties the white men acquired land titles which appeared to give them both legal rights and a quiet conscience. The empire of the Hudson's Bay Company, the frontiers of which were not clearly defined, was legally transferred from Great Britain to Canada in 1869–70. About ten years later, part of the Arctic Archipelago also passed from British hands to the new Dominion. The Sverdrup Islands only became Canadian in 1930 by a payment of $67,000. In 1927, a judgment of the Privy Council in London separated the Labrador coast from Canada, but in 1949, following a slight majority in a plebiscite, Labrador, with its guardian, the island of Newfoundland, joined Confederation.

In Canada, certain areas of water raise serious problems of ownership, or at least of jurisdiction. International law distinguishes three categories: inland waters, which are the property of the State; the territorial sea, in which jurisdiction is divided; and lastly, the high seas open to all. The Canadian waters just mentioned fall into one or another of these classes. So far as the "Canadian" sector of the Arctic is concerned, the United States, unlike the U.S.S.R., does not recognize the claims of the adjacent states. Since, in the Antarctic, the sector theory is for the moment frozen, Canada has little chance of seeing her international claim to a sector of the Arctic Ocean recognized. What then of the other marine areas around or enclosed by Canada? The width of territorial waters has been debated for a long time and the most usual limits vary between 3 and 12 miles (5.25 to 22.2 kilometres); in the latter case, the distance into bays ought not to exceed 24 miles. By taking account of the islands and of the generous 24-mile principle, the Bay of Fundy could be considered territorial despite the judgment of 1853 which virtually defined it as international. In the case of the Gulf of St. Lawrence, although the Strait of Belle Isle is narrow enough to qualify the Gulf as inland waters, Cabot Strait is a different matter. The entrance to Hudson Bay is also too large for it to be declared territorial, though historic rights (Hudson's Bay Company), declarations by Canada in 1906 and in 1937 and the fact that this sea leads to no other country all tend to favour Canadian claims. As to the channels of the Arctic Archipelago, policy has been less defined so far; moreover, according to the principle of free navigation of the seas, it would be difficult for Canada to oppose the "innocent" passage of ships on or below the surface between the Atlantic and the Pacific. It might, perhaps, be possible to establish a distinction between an "official" Northwest Passage and the other channels to be more restricted.[1] In 1970, the federal government prepared legislation called the Arctic Waters Pollution Prevention Act to protect from pollution those areas of the arctic seas which are of direct interest to Canada. Therefore, a belt of protection, 100 miles wide, has been designed around the arctic islands. Two years later, at

1. G. W. Smith, "Sovereignty in the North: The Canadian Aspect of an International Problem", in R. St. J. Macdonald, ed., *The Arctic Frontier*, pp. 194–255.

the Stockholm Conference on Human Environment, Canada played a leading role in marine pollution concerns.

It seems desirable, then, that Canada be more preoccupied with asserting her marine rights, especially where the estuaries are concerned. A policy of Canadianization would pose another problem too, that of provincialization of certain areas. There is a political dispute between the central and provincial governments concerning the jurisdiction on these marine surfaces. If provinces were to receive this responsibility, the division would not be easy in the Gulf of St. Lawrence, surrounded by five provinces.

Space and History: The Southern Frontier

Each event in history explains a bit of Canada. Here we draw attention to a few landmarks along the principal border, that of the south. Seen in an overall perspective, the influence of alternating French-English movements seems to have been more of indirect than direct impact; the fact that Quebec, Acadia, Manitoba and the Hudson Bay area were once, twice, or three times French before ultimately coming under permanent British rule has not greatly affected the extent of these areas. The territorial significance of these colonial ventures—decided in Europe and in relation to colonies elsewhere—has rather involved a component from the United States; for among the main elements which, from east to west, have created the southern cornerstones of Canadian territory is the lack of coastal American settlement beyond Maine. The thirteen colonies found themselves isolated from Nova Scotia by an unattractive band of land across which there was scarcely ever much movement. In 1763, it was in this area of Penobscot–Saint John, specifically along the Ste. Croix, that the frontier between the British colonies on the Atlantic was determined. The same reasons of noncontiguity of settlement contributed to the exclusion of Nova Scotia from the new North American republic. In 1783, and again in the treaty of 1798, the Ste. Croix was confirmed as the southeastern limit of Canada. At the same time, the famous 45th parallel was devised. A second factor causing expansion along the St. Lawrence to be limited in the south was the presence of the Iroquois on the edge of the Adirondacks, the Appalachians and even along the Windsor-Lévis axis. It is perhaps due to this intelligent and hostile tribe that settlement by New France south of Lakes Erie and Ontario was very limited and that in 1774, an immense Indian reserve was located between the thirteen colonies and the new limits of Quebec. The treaties of 1763, and others since then, have fixed the Canadian frontier south of Montreal on the 45th parallel so that one of the two metropolitan areas of Canada is only about forty miles from the United States of America.

Buffeted between Canada and her powerful neighbour, the Great Lakes region had its future decided by the trading routes between Montreal and the fur areas of the Northwest. In 1774 the extension of Quebec into the basin of the Great Lakes had not yielded to the thirteen colonies, and in the treaty of 1783 the water boundary from Lake Superior to Lake Ontario was created. The loss of considerable territory in the Michigan-Ohio area seemed to be compensated by Canadian possession of the essential trade routes for beaver pelts. Canada retained the Ottawa River, guaranteed by the treaty of 1763, and even though it gave up Michilimackinac, it retained Kaministikwia (Thunder Bay). Following the line of the Great Lakes, the southern frontier of the country edged increasingly northwards; as compared to southern Ontario it foreshadowed the latitudinal shift of the future Prairie Provinces. Worse, this evolution placed one of the forbidding areas of the Shield right in the middle of southern Canada, creating a deep structural cleavage between east and west.

In the Prairies, despite Selkirk's Red River Colony, which stretched into Minnesota and Dakota, it seems that the frontier was fixed by reference to the Lake of the Woods, on which the water trade routes from the east converged. Hence, in 1818 (ratified in 1876), the 49th parallel was chosen, a parallel which runs close to the rather unimportant watershed between the basins of the Gulf of Mexico and Hudson Bay. As a compromise, the 49th parallel was again imposed in 1846 across the Cordillera and through the Anglo-American territory of Oregon. In 1873, the Treaty of Washington extended this frontier as far as the Strait of Georgia, with both Canada and the United States finding good port sites.

Thus fixed, the southern frontier was recognized in 1908 by a treaty between the United States and the United Kingdom. It made Canada a northern country contiguous to the United States. Hence, a mixture of physical factors (Newfoundland, the St. Lawrence and its tributaries, the Great Lakes, the areas of difficult terrain, Vancouver Island) and some human factors (the French-English rivalry, the Iroquois, the beaver economy, Anglo-American interests) combined to determine the southern limits of Canada.

Internal Boundaries

The development of the political divisions within Canada was not straightforward. Take Labrador as an example—a country successively tossed between the English and the French, between Great Britain and Canada and then between Newfoundland and Quebec.[2]

Politically, Canada comprises ten provinces and two territories (fig. 8.1). The terminology *provincial* and *territorial* was extended in the nineteenth century by the British North America Act which, in 1867, brought into Confederation the colonies of Nova Scotia, New Brunswick, Quebec and Ontario. The eastern location of the four first provinces should not obscure the transcontinental aspirations of the new country. It seemed highly desirable, confronted by the westward-directed ambitions of the United States, to unite with the Atlantic the very isolated colonies of the Red River in Manitoba (established since 1811) and of the Pacific (1843–1849). Hence, Manitoba entered Confederation in 1870, and British Columbia followed in 1871. Canada at that time was extremely fragmented. An unorganized territory 300 miles (480 kilometres) wide separated Ontario from the Red River; another 700 miles (1,120 kilometres) with little or no government extended from Manitoba to British Columbia; in areal terms, Confederation was not achieved through contiguity. In 1873, it was the turn of the miniature Prince Edward Island at the other end of Canada to join this collection. This sporadic assemblage prompted the geographer Elisée Reclus to refer to Canada as a "very tenuous political entity" (1890).

During the next forty years, both the unorganized southern areas and the lands immediately to the north of the provinces were stirred to political development. In the former case, the Districts of Assiniboia, Saskatchewan, Alberta and Athabasca were created in 1882; in 1889, Ontario was extended as far west as the Lake of the Woods. So far as the northward extension of the provinces is concerned, it must be remembered that Quebec, Ontario and Manitoba did not enter Confederation in their present shape. In 1867, Quebec and Ontario had as their northern boundary the watershed determined in 1774; this represented the indistinct southern limit of the territory of the Hudson's Bay Company charter of 1670. Extension of this boundary northwards took

2. See H. Dorion, *La Frontière Québec–Terre-Neuve.*

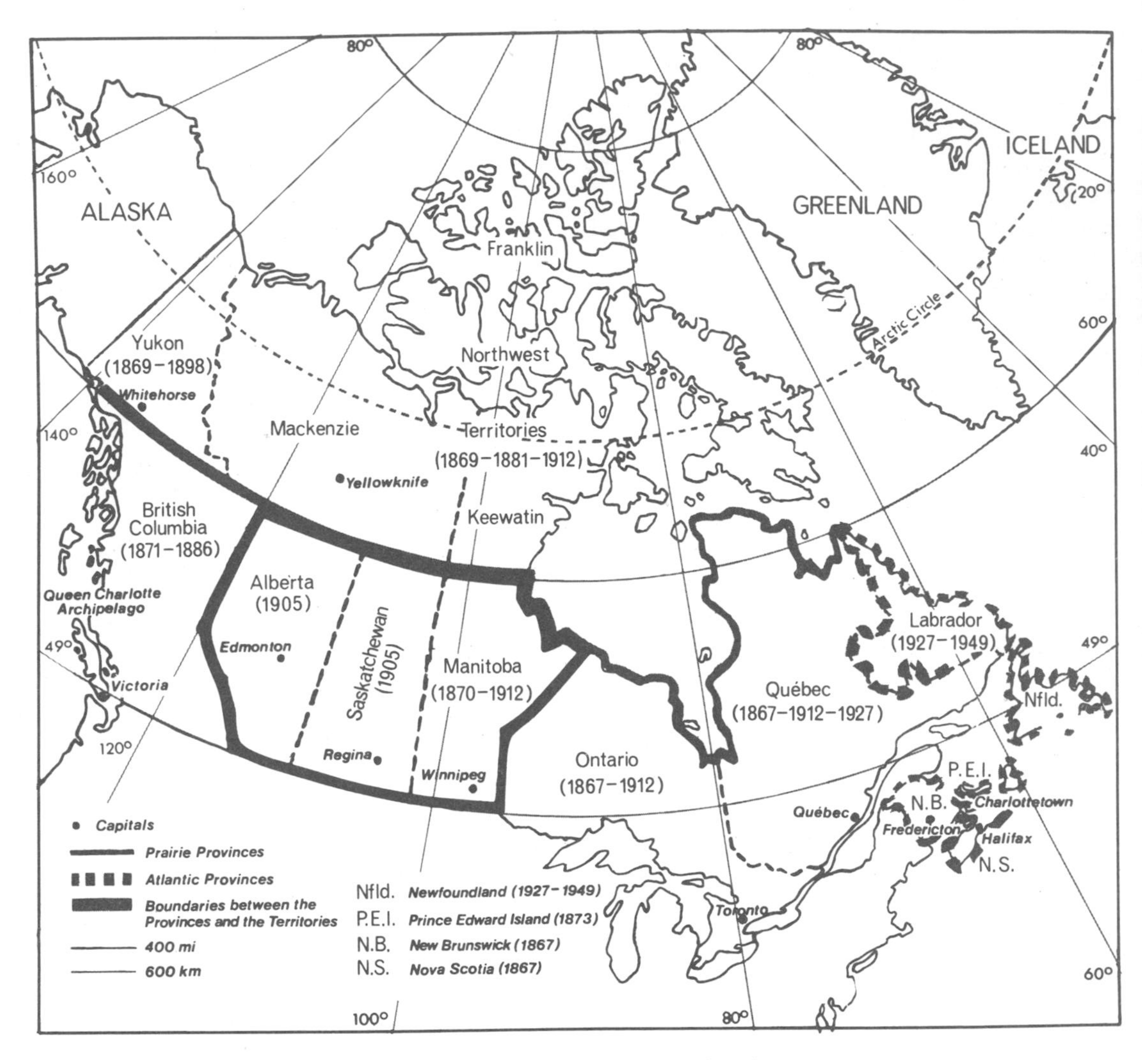

Fig. 8.1 Confederated Canada since 1867 (with main historical dates).

place gradually. Ontario reached James Bay in 1889, while Quebec got there in 1898; in 1912, Ontario reached Hudson Bay, and Quebec got as far as Hudson Strait. Quebec thus acquired the greatest northern areas and it and Newfoundland (since 1949) are the only provinces extending north of the 60th parallel. Meanwhile the Prairie Provinces also extended northwards. In 1876, the vast District of Keewatin was created, the same area which, to the displeasure of Ontario, little Manitoba acquired in 1880. Thirty-two years later the latter province reached Hudson Bay, simultaneously with Ontario. Farther west, the four districts of 1882 became, in 1905, the two provinces of Saskatchewan and Alberta. As a result of geometry, the northern frontier of the Prairie Provinces became the 60th parallel; this had been first utilized in 1866 to define British Columbia. From Alaska to Hudson Bay, there extended the longest straight-line internal boundary in the country (1,500 miles or 2,400 kilometres). The southern half of Canada was then provincialized from Atlantic (excepting Newfoundland) to Pacific.

As for the ill-defined areas north of the 60th parallel (and north of Quebec-Labrador), these eventually became two territories: the Yukon, formed in 1898 (mining economy), and the Northwest Territories. The latter was subdivided in 1912 into three districts, Mackenzie in the west, Keewatin in the east and Franklin in the north. However, more important than these districts are the four administrative areas and the ten constituencies (1970) of the government of the Northwest Territories.

The last major change in internal political units concerned the easternmost territories of Canada (those nearest to Great Britain; i.e., the island of Newfoundland and Labrador). In 1927 the "two" Labradors, coastal and interior, were, by a decision taken in London, attached to Newfoundland. Twenty-two years later, Newfoundland entered Confederation as the tenth province. Since then, in legal terms, all lands of Canada have been Canadian.

Has provincial evolution come to an end? Here, two aspects have to be considered, the first of which is the number and extent of units. There is talk of a new unit which would encompass, like the District of Columbia in the United States, the federal capital region around Ottawa, at present partly in Ontario, partly in Quebec. Elsewhere, the continuous development of the North could result in the Mackenzie region being raised from the status of a district to that of a territory, like its neighbour, the Yukon. A few years ago also, people living in the Abitibi–James Bay area either in Ontario or in Quebec were seeking to be a province. In fact, neither of the last two propositions has a chance of being realized. The regrouping of the existing provinces of Manitoba, Saskatchewan and Alberta into "one Prairie Province"[3] and Nova Scotia, New Brunswick and Prince Edward Island into one Maritime Province is more likely to be acceptable, but it is not for tomorrow. Counter to such trends, the whole provincial formula would be put into question by a total or partial withdrawal of Quebec from Confederation.

Concerning the provincial or the territorial evolution of Canada, the second factor—other than space—deals with the political structure itself. The Constitution of 1867 created a frame characterized by a large number of governmental institutions. At the present time, Canada, with its central government, ten provincial governments, two territorial "governments", and numerous town and city councils might seem to be overequipped administratively. In many areas, such as the Northwest Territories, there are parallel and conflicting exercises of power.

The British North America Act, which created Confederation, provided a complex separation of powers between the federal government and the provinces.[4] The Dominion would legislate on general matters of customs, money, credit, external trade, defence, railways and canals and criminal law, while the provinces would be responsible for, among other things, education, civil rights, municipal affairs and natural resources. This has been summarized by some as "Canadian matters by Ottawa, regional matters by the provinces".

Several fields of action which have now become of great importance were left vague. The taxing powers have been divided between central and regional governments and the latter, so as to "provide for the real needs of their inhabitants", must receive federal government grants which are proportional to the number of their inhabitants. Initial difficulties in dividing

3. See D. K. Elton, *One Prairie Province?*
4. See E. A. Driedger, *A Consolidation of the British North America Acts 1867–1965*, a report by the Department of Justice (Ottawa, 1967).

legislative competence have increased subsequently because of the different pace of development in the various regions of Canada and the increase in the number of areas in which government is active, such as social services and unemployment. The 1867 Constitution has itself been amended many times. It is still inadequate. In 1949, the entry of Newfoundland required some unusual provisions and the Northwest Territories could become a province only if granted very special conditions. In short, the authority exercised by the different governments in Canada is limited by the system of divided powers, and it is necessary either to provide coordination between the different governmental institutions or to make major changes in the structure of the country.

Federation has succeeded in appearing to be a formula adapted to the vast size of the country, but it has, in fact, effected a permanent divorce between the forces of convergence, tending towards a Canadian identity which is still to be defined, and the forces of differentiation, which are expressed by the various types of provincial autonomy. In a way more pragmatic than deliberate, and without adequate guides as to the direction to follow, the federal system is being laboriously reconstructed.

Local and Regional Government

Those who lack personal contact with life in the regions of Canada have difficulty in understanding the complicated pattern of local administration. The confusion is due to terminology, to the overlap of jurisdictions, and to the number and variety of regional institutions. Because the ecumene is very discontinuous, organization of the whole area has not been necessary and large parts in between have remained without municipal institutions. Different landscapes call for different treatment; it would not be appropriate to provide the same pattern of local government in the scattered settlement pattern of Newfoundland as in the Niagara Peninsula. The period and type of the original settlement also contributes to the variety of local institutions; Quebec in particular is very distinctive. Lastly, since the 1867 Constitution gave to each of the ten provinces individually the task of municipal organization, one could scarcely expect uniformity among the forms of government which were devised.

The regional patterns thus contain a multiplicity of elements. As recently as 1959, Quebec was divided into 78 federal electoral districts and 93 provincial ones, into 24 divisions of the provincial Legislative Council and of the federal Senate, into 30 judicial districts, 79 registration divisions, 50 cities, 154 towns, and 75 county units, which were themselves subdivided into 1,458 local units of villages, parishes and townships. These divisions still do not take into account others such as dioceses and other religious units. The situation has not improved, since this profusion is general; in all the provinces, the number and boundaries of statistical, electoral, municipal and economic regions differ from one another. The absence of uniformity can be explained in part by special needs but it is also due to the creation of separate government departments and the lack of administrative planning.

The structures of counties and of municipalities should be simplified, made more uniform and more adjusted to geographical reality; it is particularly the suburban areas and those on the rural margin which are most in need of reform. No territorial modification can be lasting, however, without a new division of taxing powers between federal, provincial and municipal governments. It is not healthy for the municipalities to be, as they are, dependent on government departments. Paradoxically too, while local administrative units seem to be overly complex, some sociologists, like F. Dumont, argue that the development of rural areas is hindered by a lack of structure.[5]

5. See F. Dumont and Y. Martin, *L'analyse des structures sociales régionales* (Quebec, 1963).

Chapter 9

Regional Disparities

The main characteristic of the geography of Canada lies in the differences between regions; these contrasts concern many, if not all, aspects of a country: physical components, resources, people, economic activity, cultural life. There are disparities not only from one group of provinces to another group, but also inside the same province. These contrasts are so great that they produce one of the biggest problems Canada has to solve.

Contrasts Between Major Regions

In a young country with a discontinuous ecumene like Canada, European notions of what constitutes a region only apply to a small part of the land area. The reader must therefore not expect to find here a regional study of the French type. On the contrary, the natural landscape, historical events and present-day political divisions have given rise to sharp economic contrasts in Canada. Five areas, which might be called *megaprovinces*, can be identified: the four Atlantic Provinces, Quebec, Ontario, the three provinces of Alberta, Saskatchewan and Manitoba, and British Columbia. North of these provinces are the Yukon Territory and the Northwest Territories.

A General Perspective

To appreciate the real geographical differences between the provinces, a more comprehensive view is required than that which is suggested by economic facts alone. Following Griffith Taylor's notion of a "dynograph",[1] we can compare the megaprovinces on the basis of eight major characteristics: total area, area of the ecumene, population, personal revenue, manufacturing output, agricultural output, mining and fuels activity and hydroelectricity generation. Graphically, the regions form octagons within a basic circle (fig. 9.1). For statistical and geographical reasons, Ontario has been chosen as the standard. In this province, therefore, each of the elements considered is assigned the fixed value of 20, and values in terms of this figure are calculated for the other megaprovinces. For example, for 1971 the 7,703,106 inhabitants of Ontario were defined as 20; the 6,027,764 inhabitants of Quebec thus yielded a coefficient of 15.

Ontario is the dominant region in almost all the indices; most sides of the octagons of the other regions lie within the circles. Only seven points out of a possible forty are outside the circles and three of these represent area (the Territories, Quebec, and the Prairies are each larger than Ontario); three other exceptions are related to the area of the ecumene, to agricultural output and to mining. These occur in the Prairies, which have the most irregular polygon. The seventh exception concerns hydroelectricity in Quebec. British Columbia—a province which is not grouped with others, as are its three neighbours to the east—appears weak in agriculture and manufacturing, but fairly rich in area and in energy (and also in forest resources, which are not included here). The Atlantic region shows, at least when compared to Ontario, a contrast between the area of its ecumene and the value of its agricultural output. Manufacturing development is also low, despite the long history of settlement of the region. Quebec is reasonably similar to Ontario in terms of ecumene, population and energy production, but differs from Ontario regarding contribution to personal revenue and manufacturing output, and only provides half the Ontarian proportion of mining and agriculture.

For a general view of Canadian regional disparities, see D. M. Ray, *Dimensions of Canadian Regionalism*, Geographical Paper, No. 49 (Ottawa: Information Canada, 1971).

1. G. Taylor, *Canada. A Study of Cool, Continental Environments and Their Effect on British and French Settlement*, p. 490.

A general comparison between the individual values for population and for contribution to personal revenue shows that productivity is low in the Atlantic area; at the level of the megaprovinces only British Columbia has a productivity level comparable to that of Ontario. The ratio of area of ecumene to number of inhabitants shows that real population densities gradually diminish from Ontario through Quebec, British Columbia, the Atlantic Provinces and the Prairies to the Territories. Population movements (immigration to Ontario, emigration from the Atlantic Provinces) and output (agriculture in the Prairies) are among the factors which explain why the hierarchy of densities does not completely coincide with the length of time since each region was settled. With a few exceptions, therefore, Ontario is the dominant, or at least the most significant, region.

Structural Elements

The thousands of provincial disparities, as well as the fundamental and quasi-permanent characteristics of the regional economies, owe their origin to basic facts, the majority of which have a recognizable geographical significance.

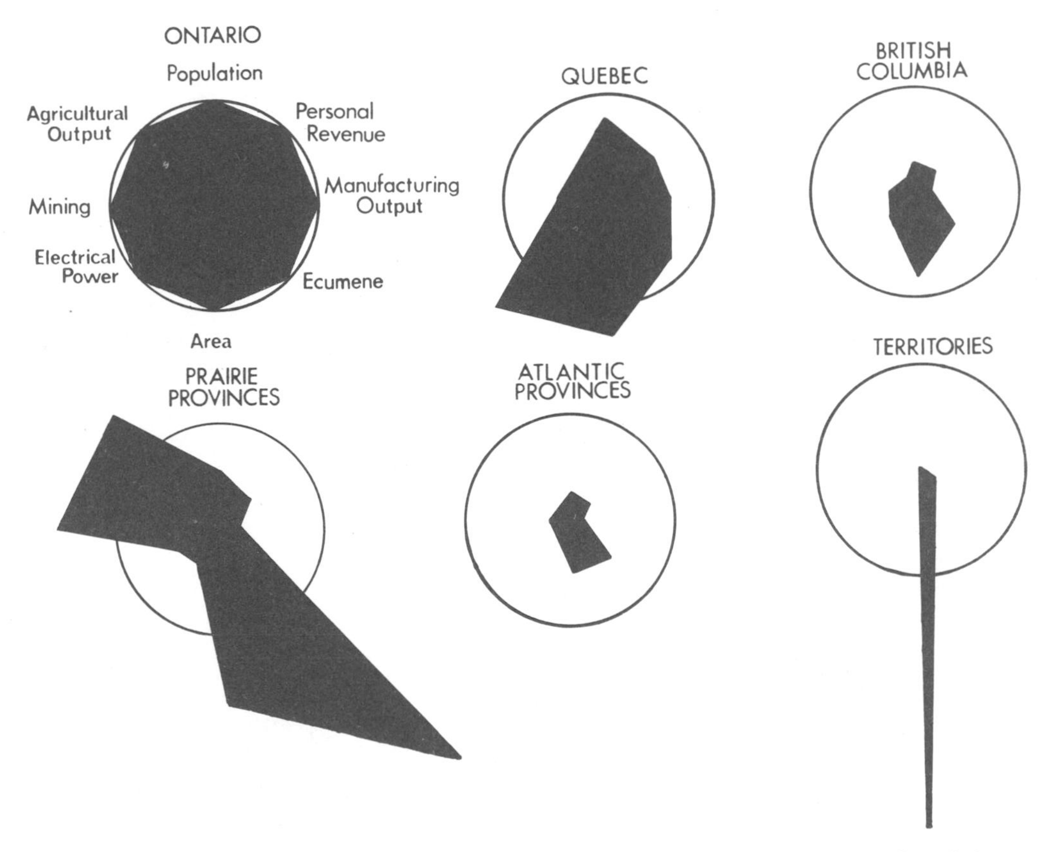

Fig. 9.1 Comparative profile of the megaprovinces, 1970. (Data from Information Canada.)

Ontario

In Canada, Ontario is a privileged place; well-endowed by nature, its advantages have increased during the short two centuries of its history. Possessing 35 per cent of the population of the country, providing 40 per cent of the GNP and more than 50 per cent of the manufacturing activity, economic Ontario sets the pace for the other provinces, while cultural Ontario dominates English-speaking Canada. Even though provincial government and federal government must always be distinguished, the political capital of Canada, Ottawa, is also located in this major province. Though the per capita income of Ontario is below the United States average, it is well above many parts of that country. In economic terms, Ontario in Canada is in a similar position to the state of São Paulo in Brazil.

Indisputably, among the favourable factors is Ontario's location: an interlake spur reaching into one of the most advanced economic regions in the world. This nearness to the United States has a direct effect on population movements, the supply of raw materials, markets, the spill-over of manufacturing across the narrow divides at Niagara, Windsor and Sarnia and the flow of culture (e.g., television). Contiguity with the United States, instead of creating a competitive situation, has tended to lead to complementary forms of activity.

Other favourable factors in the site include the opportunities for shipping; although far from the open sea, Ontario has a gateway about 1,000 miles (1,600 kilometres) wide on to the Great Lakes, enabling low-cost transport of heavy products, an attraction to industry. Ontario has also benefited, since the days of canals, from waterways to Atlantic ports in both Canada and the United States, allowing the greater part of the province to be linked with eastern Canada. Ontario's advantages include a southerly position permitting rich farming and posing fewer problems in the Canadian winter. Present-day advantages also include an accumulation of labour and of spending power unequalled in Canada. This concentration of population in a part of southern Ontario which does not amount to 1 per cent of the national land area automatically creates the best domestic market. Modern Ontario has provided itself with a powerful infrastructure centred on Toronto and extending nowadays in a horseshoe-shaped region more than 100 miles (160 kilometres) from Oshawa to St. Catharines. In relation to the Prairies, Ontario has been both the entry route of immigrants and the means by which their products are exported. Cereals, animal products, oil and gas form the chief imports, and are exchanged for manufactured articles and services. Lastly, the Ontarian economy is full of dynamism; along the metalliferous edge of the Shield, closely watched by the Toronto money market, new enterprises replace the gold and uranium mines which are now in difficulty. Since 1963, investments have grown rapidly in the secondary sector, which is becoming diversified. During the 1965–66 winter, the unemployment rate did not exceed 2 per cent. Since 1966, Douglas Point on Lake Huron has produced nuclear energy. Ontario has developed thermal and nuclear-produced electricity, in contrast to Quebec and British Columbia. It is increasingly accepted that suburban growth and the decay of town centres impose a need for planning. Toronto is transforming itself. In this province, central by virtue of its influence rather than its geometric position, favourable factors from all the available evidence are more important than the disadvantages. It is not easy to see how Ontario could lose its position of leadership.

Quebec

Quebec, the only other major province, contains one-quarter (28 per cent) of the population and of Canadian production. It possesses the most important urban concentration in Canada and, with Toronto, Canada's only other major international city, Montreal. Quebec leads in the production of those staples which, throughout North America, have built so many fortunes. Even from an economic viewpoint, it is important to note that Quebec cannot be evaluated without taking into account its chief cultural characteristics. By virtue of its age and its French culture, this hub of a former empire has caused Canada to be a very different country from that which would have resulted from a federation only of British colonies. Contemporary feelings in Quebec, where the French-speaking population represents about 80 per cent of the total and dominates the political life of the province, are based partly on the economic poverty of French Canadians, by comparison with their English-speaking fellow citizens. In Ontario, economic matters do not have a similar cultural significance.

Quebec also has great natural resources. First is the St. Lawrence, along which the first permanent settlements developed. The St. Lawrence is the gateway to Europe, marked by ports stretching from Quebec City upstream to Montreal and downstream to Baie-Comeau, Port Cartier and Sept-Iles. The St. Lawrence is the best Canadian outlet from the vast continental interior. Once from that interior came furs carried by canoes; today it is grain or ore carried by rail. The St. Lawrence is also the best way of bringing in the raw materials for Montreal industry. The river, however, did not facilitate links between its different parts. Until the end of 1967, only a single bridge crossed the river below Montreal. Consequently, southern Quebec is a double region which has developed linearly along the north and south shores of the St. Lawrence. The small-scale character of agriculture, climatic problems, and very belated industrialization have led to heavy emigration, a safety valve for a high rate of natural increase. Spatially, Quebec is very close to the Shield, a rocky and wet Shield which has repelled agriculture despite the best efforts of the province, but which supports one of the most productive forests (in terms of paper) in the world and also the development of metallic minerals, including iron mining (since 1954). These raw materials have attracted capital and industrial magnates, first from Great Britain and then from the United States; this foreign assistance has enabled the growth of rich oases remote from the main centres, as at Schefferville and Chibougamau.

Organized Quebec, despite colonizing ventures northward, has still not reached the lower North Shore or the shores of James Bay. In the vast area of Quebec, the main economic life extends very little beyond the tiny region of the St. Lawrence Lowlands. Dynamic Quebec is increasingly becoming Greater Montreal, which has benefited in recent years from the enormous social investment associated with the holding of the 1967 International Exposition (Expo '67). The contrast is enormous between booming Montreal and the underdeveloped Gaspé, and even more obvious with the emptiness of the Quebec-Labrador peninsula. The *Québec du Moyen-Estuaire* (downstream from Quebec City) has the lowest personal income in the whole of Canada; this broad area of the St. Lawrence is much more like the Maritimes than it is like the Montreal area.

It is not yet possible to foresee all the consequences of the new awareness which has characterized the last fifteen years in Quebec. It was not sparked by investors as is the normal case in North America; it came from the people and from government. The latter has begun or has revived a whole series of measures for economic and cultural renewal: public ownership of electricity, the Bureau for the Development of the East of Quebec (1963), the Office of French Language, the General Investment Corporation, the Quebec Mining Exploration Corporation (1965), and the Ministry of Education.

Plate 9.1 From the hills of the Gaspé Peninsula, a view southwards of Carleton-sur-Mer, the Baie des Chaleurs, and northern New Brunswick. The typical linear field patterns stretch from the woodlot down to the line of farmsteads along the *rang*. This region has one of the lowest average per capita incomes in Canada; emigration and land abandonment are common.

Even though there is a serious shortage of senior civil servants able to implement these initiatives, they represent a clear desire to use political power in Quebec for the benefit of the French-Canadian community (see chapter 12).

The Prairie Provinces

Unlike Quebec, and despite the polyethnic mixture of these provinces, the economic life of the Prairies has been almost a matter of business, or rather of big business; it was little influenced by cultural considerations or even by economic nationalism. Although the Prairie Provinces have traditionally been considered a single unit, it is becoming increasingly necessary to separate Alberta from Manitoba and Saskatchewan. In Alberta, per capita investment is now the highest in the country; in 1965, Alberta generated 44 per cent of the personal income of the three provinces. Taken together, the Prairies dominate Canada in terms of area of ecumene

(Saskatchewan has 40 per cent of the cleared land area of Canada), of agricultural production and, for the last twenty-five years, of fuel production. It is an immense region, the settlement of which gave a new look to Canadian international trade. By becoming aware of their own identity, the Prairies have created one of the most powerful manifestations of a type of Canadianization in which the concept of two nations "is wrong".[2] In addition, they now protest against their apparent semicolonial domination by eastern Canada.

The Prairie Provinces lie beyond the barrier of the Shield in northwestern Ontario, east of the Rockies, and are open both towards the south (U.S.A.) and the north (Churchill and the Mackenzie). The factors which aid the subcontinental economy rest on a collection of edaphic, climatic, biological, technical and commercial characteristics which enable the widespread and profitable cultivation of cereals. Because wheat is harvested only at the end of the summer, an enormous network of railways, large-capacity elevator systems and a specialized fleet of lakers had to be provided to ship the crop before the freeze-up of the Great Lakes and the still earlier freezing of Hudson Bay. Movement across the Cordillera to the Pacific Ocean is less of a problem. For several decades, a substantial fall in size of the agricultural population, compensated by mechanization and the development of highway transport, has led to the restructuring of the railway net and the methods of storing grain. Since grain has remained the basis of the economy, at least in Saskatchewan, the prosperity of the Prairies is dependent on climatic hazards, seasonal unemployment and on supply and demand in the world market. Some years in the last decennial period have been very prosperous as a result of abundant crops (wheat) and particularly because of great demands from socialist countries (the U.S.S.R. and China).

Fuel resources, the second prop of the Prairie economy, are also activities in the primary sector. There have been problems both of transport (oil and gas pipelines) and of markets: Ontario, Montreal and Vancouver. Easy access to the U.S. Midwest and to California is one of the essential requirements for massive growth in production of petroleum and gas. Arctic fuels can reach southern markets through Alberta.

The development of industry in the Prairie Provinces is not helped by the restricted size of the regional market or by the rivalry of Ontario. However, industries based on local raw materials such as petrochemicals and potash have recently made spectacular developments.

Lastly, the Prairie Provinces include a portion of the Shield. Although, until now, it has been less exploited here than in Ontario and Quebec, there are nevertheless about ten mining outposts, including Thompson (nickel). The Nelson is being harnessed for hydroelectric power, and the Prairies are linked across the Shield with the growing export port of Churchill.

Although the Prairies are still a producer of raw materials, there has been a tremendous urban development; before the war, Winnipeg was the only big city. Since then, Edmonton and Calgary have become among the fastest growing agglomerations in Canada.[3]

British Columbia

Canadian Columbia, as Pacific and as closely related to the U.S.A. as it is "British", is the only province other than Ontario and Quebec to be considered by itself. This distinction is related less to the absolute size of its output and

2. J. Diefenbaker, Progressive Conservative Party National Convention, 1967.

3. B. Proudfoot, "Wheat Harvest Moon Wanes over the Prairies", *The Geographical Magazine* XLIV (1972): 385–392.

its population (about 10 per cent of the national total) than to its individuality.

Its advantages are undoubtedly the presence of abundant raw materials: mines, fisheries and especially forest resources. These directly or indirectly provide half the total income of the province. The forests are a response to the plentiful winter precipitation combined with a long growing season. Secondly, British Columbia has a mountainous landscape in which the ecumene, apart from scattered or linear patterns, is limited to the periphery or the bottom of the valleys. This has been one factor in the growth of the major urban region around the Strait of Georgia. The relief also raises the costs of intraprovincial transport and, until World War II, made interregional links between the coast and the interior difficult.

British Columbia is a Pacific region which was linked to Great Britain long before it developed links with eastern Canada. Only with the coming of the railway in 1886 were the British colonies in the far west connected to the distant St. Lawrence. (Vancouver was chosen as the site of the rail terminus because of its potential as a port). The opening of the Panama Canal in 1914, the progressive development of relations with the coast of the United States, the war and then the trade with Japan (the Mitsui group), and finally the increasing westward movement of Prairie products have all enlarged the Pacific outlook of British Columbia.

Greater Vancouver, the third-largest population group in Canada, dominates the provincial economy in the same way that Winnipeg does in Manitoba, Montreal in Quebec and Toronto in Ontario. It is the second port and has recently become the second financial centre of the country. Since there is a direct relationship between growth areas and income, the rapid urbanization of the province is accompanied by high wages (in the summer of 1966, industrial salaries were up to 8 per cent higher than those in Ontario) and by high levels of productivity per worker. Such conditions have been an attraction to easterners.

As in Quebec, provincial political power has become a factor in economic development. This is exemplified in the fantastic hydroelectric projects on the Columbia and the Peace rivers and by a hard line with an Ottawa which might otherwise tend to ignore the distinctive problems of the remote west coast. As elsewhere in Canada, the relative buoyancy of external markets has direct effects on provincial economic prosperity.

The Atlantic Provinces

What a contrast there is in age, power and standard of living between the Atlantic and Pacific coasts. The three Maritime Provinces and Newfoundland-Labrador are the main underdeveloped regions in southern Canada. Average wage income is 20 per cent below the Canadian average, whereas in British Columbia it is 14 per cent above. Containing 10 per cent of the national population, these four provinces provide only 5 per cent of the total national income, which implies a low level of productivity.

For more than a century, and particularly since World War II, the area's disadvantages have appeared dominant. The Maritimes, which provided 70 per cent of the Fathers of Confederation, today have only 12 per cent of the members in the federal parliament. The loss of any real advantage in closeness to Europe, due to the development of modern communications and changes in trade routes; land fragmentation, which is accentuated by continuing transport difficulties and by the existence of separate political units within the region; the cultural separation of the English-speaking part of the Maritimes from the rest of English Canada by Quebec and northern New Brunswick; a location outside the heartland of North America, which, though it does not protect the Maritimes

Plate 9.2 The older sections of St. John's, Nfld., suggest the influence of continental Europe, despite frame construction and the overwhelmingly British Isles origins of the population. Redundant chimneys provide support for television aerials. The photograph was taken in 1964.

from heartland competition, does prevent reciprocal access to the main continental markets; the absence of a major metropolitan area in an age of large cities—all these tend to make the Atlantic region, grouped around that virtually Canadian sea, the Gulf of St. Lawrence, seem no more than the historic outport of Canada. It is true that there are some major new enterprises, such as the hydroelectric dams of Mataquac in New Brunswick and Baie d'Espoir and Churchill Falls in Newfoundland, heavy water plants in Nova Scotia, and an industrial complex near Bathurst. These developments are not sufficient, however, to stop the emigration which deprives the region of producers, consumers and ideas; such projects do not provide a rate of growth sufficient for regional employment needs. If the cultural destiny of Confederation is being worked out in Quebec, its economic destiny is a matter for the Atlantic Provinces. Can a central government succeed in checking the natural tendency towards decline? There do not seem to be any simple or quick solutions.

Contrasts Within Provinces

From one province to another, economic activity is very far from being uniformly distributed or composed of similar elements.

Types of Activities

The three principal sources of income in terms of *value added* in production are presented in table 9.1, by percentages in the major regions.

(1) In Ontario (manufacturing), Quebec (manufacturing) and the Territories (mining), one sector clearly dominates; these megaunits seem overspecialized.

(2) In four provinces—British Columbia (manufacturing), New Brunswick (manufacturing), Nova Scotia (manufacturing) and Saskatchewan (agriculture)—the dominant category does not reach the 50 per cent mark but is over 40 per cent.

(3) Lastly, in Prince Edward Island (agriculture), Alberta (mines), Newfoundland-Labrador (mines) and Manitoba (manufacturing), because the contribution of the principal sector is less, the forms of economic activity seem more in balance.

During recent years, the Prairie Provinces have recorded the greatest changes with an increase in mining activities and a relative decrease in agricultural value.

Spatial Contrasts

On the basis of *market index* (per capita disposable income in each region compared to that of Canada as a whole),[4] the lowest values have been subtracted from the highest values within each province, except for Prince Edward Island, which has been considered as a whole (table 9.2). In general, the metropolitan areas and the northern mining outposts of activity have very high incomes, while the marginal fringes of old rural settlement have low ones. In 1961, the province which had the greatest internal economic differences was Quebec. There were in fact two marked extremes: Montreal and the Gaspé–South Shore region.[5] Even if Montreal is excluded, there remain large differences between the other industrialized regions and the Appalachian sector of Quebec along the estuary. The latter had the lowest index (a value of 46) of the sixty-eight economic regions of Canada (see p. 85). This is naturally a region whose future was of concern to A.R.D.A.* Manitoba also showed strong intraregional variation: there were great contrasts in income between the northern mining outposts such as Flin Flon, and the zone along the frontier of continuous settlement, like the Interlake. Next, somewhat surprisingly, comes Ontario, where the contrasts are between Toronto and the inhabited edge of the Shield. Newfoundland has a similar variation between the high incomes of the mining areas of Labrador and the contrasting Avalon Peninsula, where the provincial capital of St. John's is located. In fifth place is British Columbia, where there are major differences between the urbanized region on the coast and the Okanagan interior.

Two other comments can be made about these disparities. The first concerns their tendency to change. Whether at the level of major regions or at the level of intraprovincial units, the contrasts have an unhappy tendency to become more marked, despite the policies and programs of governments. The economic contrast between the southeast and the southwest

4. See P. Camu, E. P. Weeks and Z. W. Sametz, *Economic Geography of Canada.*

5. See [Canada], *Development Plan for the Pilot Region: Lower St. Lawrence, Gaspé and Iles-de-la-Madeleine. A Summary*, a report by the Department of Forestry and Rural Development (Ottawa, 1967).

* The Agricultural and Rural Development Act, 1966, or the Agricultural Rehabilitation and Development Act, 1961.

TABLE 9.1 Value Added (%) for Goods-Producing Industries, 1968

	Agricul-ture	*Forestry*	*Fisheries, Trapping*	*Mining*	*Electric Power*	*Manufac-turing*	*Construc-tion*	*Total**
Atlantic Provinces				14		37	26	77
Quebec				5.7		67.9	13.9	87.5
Ontario	5.7					71	13.6	90.3
Prairie Provinces	26.6			25.5		21	22	74.1
British Columbia		11.4				48.5	21.3	81.2
Territories				88.2	7.1	2.6		97.9
Canada	9			9.9		57.4	16.5	83.8

* Three most important sectors.

SOURCE: *Canada Year Book, 1971.*

TABLE 9.2 Income Contrasts Within Provinces, 1961

	Contrast of Extremes	*Provincial Ranking by Average Per Capita Income*	*Provincial Ranking by Urbanized Proportion of Population*
Quebec	68	5	2
Manitoba	56	4	4
Ontario	46	2	1
Newfoundland	44	10	7
British Columbia	42	1	3
Alberta	35	3	5
Saskatchewan	30	7	9
Nova Scotia	19	6	6
New Brunswick	16	8	8
Prince Edward Island	—	9	10

SOURCE: Calculated from data in P. Camu, E. P. Weeks, and Z. W. Sametz, *Economic Geography of Canada.*

of Canada is increasing. In southern Canada, all areas east of eastern Ontario (except for Greater Montreal) have a per capita disposable income lower than in the poorest part of British Columbia. Secondly, even though the greatest internal inequalities occur in the poorer provinces, they are more significant in those provinces which are moderately endowed but also highly urbanized, for example, Quebec and Manitoba. The provinces which do not have major metropolitan areas are less likely to have major internal differences; they include New Brunswick, Nova Scotia and even Saskatchewan. Economic disparities do not hit only the southern facade of the country, but thrust also into the Canadian North. A spatial contrast is growing now between the Mackenzie and Keewatin regions.

The Choice of Development Areas

In June 1961, the Agricultural Rehabilitation and Development Act was intended to lessen the poverty of rural areas. The areas concerned were identified by a low index of output, high levels of unemployment and underemployment, low capital investment, low quality of housing, low levels of education and technical training, and finally by low tax revenues. Such areas are a national liability: they receive more from various governments than they contribute to the growth in Gross National Product. Their unfortunate condition might seem scandalous when compared to the abundance in other areas. In contrast to the United States, poverty is not linked to racial segregation, though it is more significant among the native peoples, French Canadians and non-British immigrants.

In Canada, studies of rural underdevelopment have generated a variety of solutions, some of which are relevant to the overall condition and others only to local problems. Some prescriptions for improvement, such as the development of staples and the free movement of both capital and workers, reflect the durable American concept of a free economy. Such solutions generate more problems because, in the first place, production of staples does not lead to all-around development and, secondly, economic migration in search of rapid growth will rarely completely empty an area of its inhabitants. Even if the latter were to happen, there would remain behind an impoverished landscape. By contrast, federal officials utilize the opportunities of central government, through fiscal agreements, transfer payments (from federal to provincial and municipal governments), special subsidies to transport links, direct public investments (as in Prince Edward Island) and social investments (adult education, technical training). In a variety of ways, such national help must be allied with regional efforts. At the local level, spontaneous emigration, such as has happened in the Maritimes for more than a century, does not seem to be an adequate remedy. The aim should be to develop a spirit of initiative and the ability to adapt to major change, and to create local leadership. In the have-not provinces, the vicious circle of reduced output, low profits and consequent limited investments must be broken in those activities which can become generators of further growth. The basic objective is to increase the opportunities for high productivity employment.

The concern of A.R.D.A. for the alleviation of rural misery leads one to hope for a comprehensive development policy which would also encompass the needs of suburban and urban areas. This new approach has already been put in legislation.

"In 1969, a new federal Department of Regional Economic Expansion (DREE) was created, to assume overall responsibility by the federal government for reducing regional disparities. The basic strategy of the department has been described as 'the promotion of regional growth by the encouragement of industrial

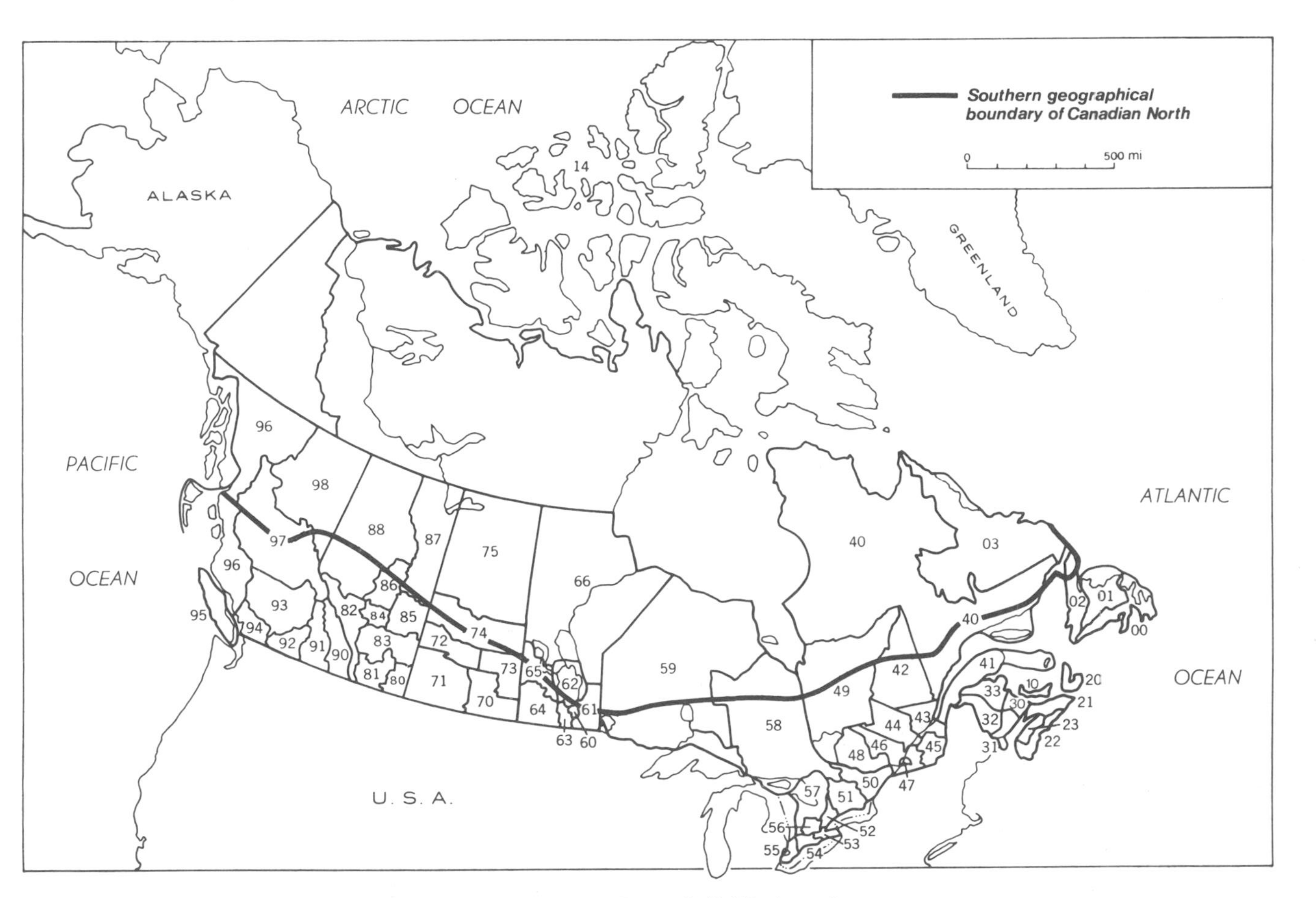

Fig. 9.2 Provincial economic regions of Canada. (From P. Camu, E. P. Weeks and Z. W. Sametz, *Economic Geography of Canada.*)

expansion in urban centres located in economically depressed areas'. The time frame envisaged as necessary for significant change is, as in the Prince Edward Island Plan, fifteen years.

"Close to the core of this approach is, of course, the idea of growth points: the concentration of effort on a relatively few areas which offer the greatest opportunities. DREE identified four such areas in the Maritimes: Moncton and Saint John in New Brunswick, Halifax-Dartmouth and the Strait of Canso in Nova Scotia. The first three of these are significant regional centres already, where 'faster growth can have major repercussions through eastern Canada generally'. The Strait of Canso is the site of major new industrial growth, due principally to the ease of access to its new refineries by supertankers. A fifth area, based on the long-established coal and steel industries of Sydney, Nova Scotia, receives special help from the federal government through the Cape Breton Development Corporation."[6]

In the war on poverty, two elements—unemployment and production—are of main significance, but the landscape as such has to be considered also; the objective is a total approach. Further planning must not limit itself to revival; it must be involved from the beginning in the initial development of frontier regions, especially in the Middle North. Lastly, to counter the diverse autonomies of individual agencies and governments, better coordination and communication must be provided so that unified efforts will bring greater rewards.

Regional Divisions

The regional division of Canada is presented here only in very general terms. A variety of criteria—structural (distribution of resources), administrative (census division boundaries) and strictly economic (spheres of influence of trade centres)—is contained in this regional division. Only the results of the analysis are presented here; comments on them will be found in other parts of this book.

Economic Regions

Following World War II and the Korean War, the federal Department of Defence Production became concerned with those regions which had problems of economic reconversion. In 1953, it designated preliminary economic regions (fig. 9.2). Mainly because of their statistical usefulness, these regional divisions were used by various agencies. They can be found, with slight modifications, in the *Resources for Tomorrow* conference in 1961,[7] and very clearly in the work of Camu, Weeks and Sametz.[8] In terms of a spatial hierarchy, these regions fall midway between reasonably homogeneous mesoregions and the megaprovinces, which contain numerous internal contrasts. According to Camu, Weeks and Sametz, the 68 principal regions contain 288 smaller functional zones, but the regions can themselves be grouped into 14 larger units (provinces or territorial districts). This assists quantitative analysis and is one of the advantages of this classification. The commercial index (market index) of each major region expresses the personal disposable income (after taxes) in each region as a percentage of the Canadian average. The key characteristics of these regions are included in table 9.3. The Territories have been excluded.

The geographical integrity of this regional division is constrained by concessions to administrative or census-division boundaries, or to decimalization (the coding system does not provide for more than ten regions per province, which is as much a nuisance for Quebec as it is for Ontario), and it is not suitable for the Canadian North. However, such a division does provide a rapid and accurate picture of the

6. M. C. Storrie, "Prognosis for Atlantic Provinces", *The Geographical Magazine* XLIV (1972): 679.

7. See [Canada], *Resources for Tomorrow*, Vol. 1 (Ottawa, 1961).

8. P. Camu, E. P. Weeks and Z. W. Sametz, *Economic Geography of Canada*.

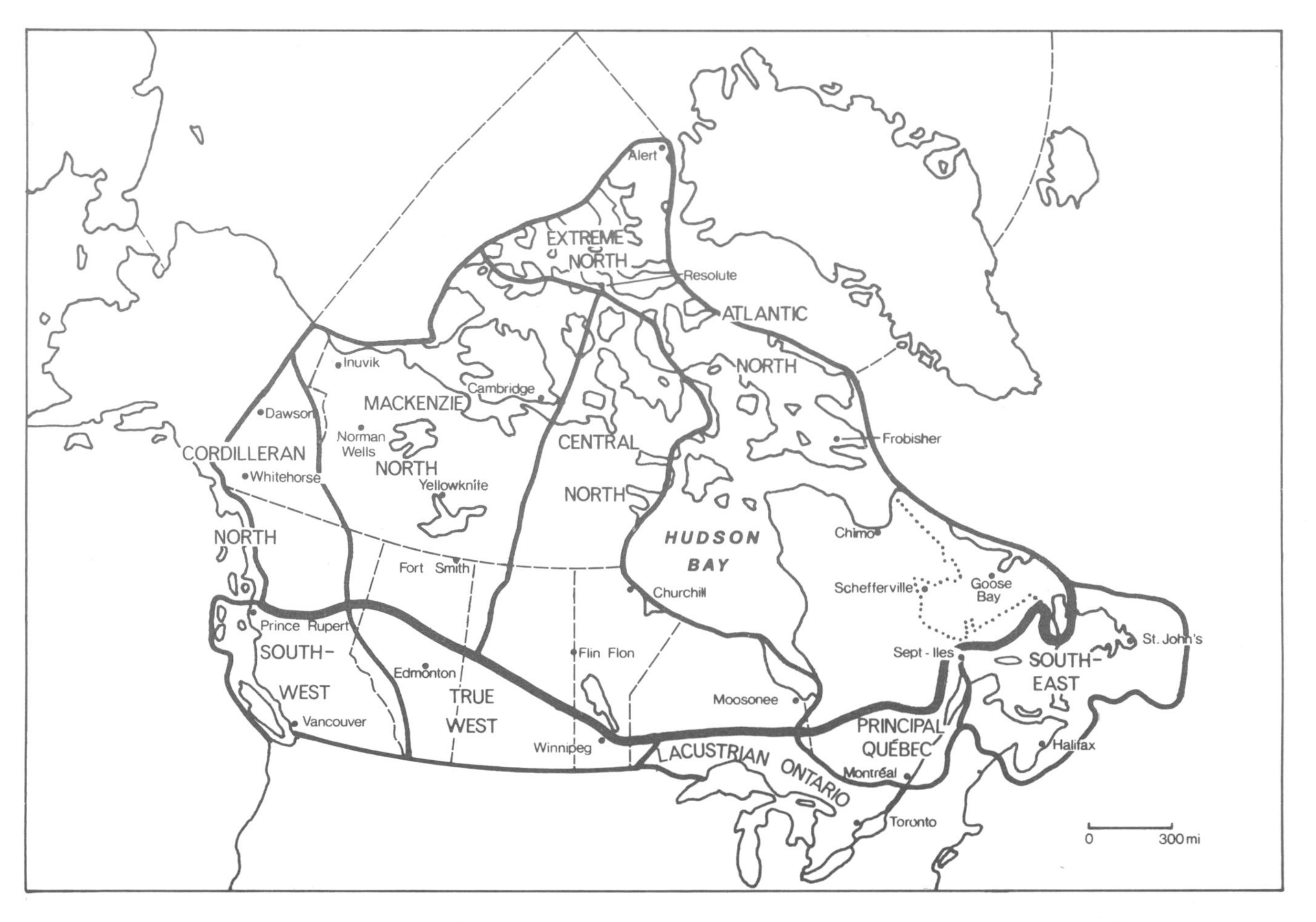

Fig. 9.3 Canadian major regions, north and south.

spatial diversification of the economy, and a panoramic view of the relative affluence of different regions. Further, the market index, when compared with the size of the population, suggests the overall importance of each region.

Major Divisions

Seen on a small-scale map, geographical Canada might comprise nine large units, five of which are situated in the southern part of the country (fig. 9.3).

Southern (Base) Canada

The megaprovinces discussed in the earlier part of this chapter have external boundaries that we have slightly modified so as to make them a better fit with geographical reality. (Southern Canada, on the basis of the polar index discussed earlier, does not include the northern parts of the seven provinces.) For example, Quebec "loses" the eastern part of its Appalachian area (South Shore–Gaspé) to the southeastern region of Canada around the Gulf of St. Lawrence and the Atlantic banks. Conversely, along its western edge, southern Quebec both gains and loses. Although its influence extends beyond rural Abitibi and into easternmost Ontario, Quebec shares leadership with Ontario in the Hull and Rouyn regions. In northwestern Ontario, the area west of Dryden and Lake of the Woods comes within the sphere of influence of Manitoba. Lastly, in landscape terms at least, the tourist areas of the Rockies are becoming more a part of the Cordillera than of the Prairies. British Columbia is also extended on its western side through the addition of channels, straits and fiords of the Pacific littoral. Due to the existence of natural breaks in the ecumene along the edges of the megaprovinces, such slight changes in boundaries modify only to a small extent the total number of inhabitants and the economic characteristics of the large provincial units. Nevertheless, because of these modifications—even though they are slight—it is preferable to avoid using a strictly provincial terminology to describe the new units. It is suggested, therefore, that these regions be called, from east to west: Southeast, Principal Quebec, Lacustrian Ontario, True West, and Southwest.*

Canadian North

In its turn, the immense Canadian North can be divided into four large zones, roughly along longitudinal lines, which are basically defined by the complex of relationships that each zone has with southern Canada. These zones cross the latitudinal Middle North and Far North.

(1) In the northwest, a *Cordilleran North* includes much of the Yukon and parts of northern British Columbia and the Alberta plateau. The Yukon River and the Alaska Highway have been and remain major elements in the regional structure. South of the old mining settlement of Dawson, the service centre of Greater Whitehorse, with 11,000 inhabitants, is one of the largest population concentrations in the entire Canadian North.

(2) To the east of the preceding region is the *Mackenzian North*, bounded on the south by the True West and on the north by the Extreme North. Yellowknife, on Great Slave Lake, is the new capital of the Northwest Territories. Along the Mackenzie axis is the most active region in the Territories.

(3) The *Central North* is very much less developed, apart from the mining areas in the Middle North and the Churchill Railway.

(4) Lastly, an *Atlantic North* with navigable sea routes stretches from Labrador to Ellesmere Island and through Hudson Bay.

* This terminology is a literal translation from the French: *Sud-Est, Québec principal, Ontario lacustre, Ouest authentique, Sud-Ouest.* Because these terms do not translate easily, the more conventional names (Atlantic region, etc.) are used in most places elsewhere in this book. Similarly, *Canada de base* is normally translated as "southern Canada". The equivalence is not, however, exact; for example, the Atlantic Provinces include Labrador, but Hamelin's *Sud-Est* does not.—*trans.*

TABLE 9.3 Intraprovincial Economic Regions

Code	Major Regions	Total Population, 1961 (thousands)	Market Index	Leading Economic Features
	Newfoundland:			
00	Eastern and southern Newfoundland	277	56	Minimum provincial index. St. John's (capital). Close to Grand Banks (fisheries).
01	Central Newfoundland	83	62	Grand Falls (paper). Exploits River basin. Trans-Canada Highway.
02	Western Newfoundland	85	73	Corner Brook (paper). Humber Basin. Port-aux-Basques (railway, port).
03	Labrador (1927)	14	100	Iron. Hydroelectric potential. Problems of the border. Economically linked with Quebec. Several settled oases.
10	*Prince Edward Island:*	105	66	Charlottetown (capital). Agriculture (potatoes). Development plan.
	Nova Scotia:			
20	Cape Breton	170	84	Peninsula (since the causeway). Coal and steel (Sydney).
21	Central Nova Scotia	143	68	Minimum index. Truro. Pictou.
22	Southern Nova Scotia	332	87	Halifax (capital, year-round port).
23	Annapolis Valley	90	72	Agriculture and tourism.
	New Brunswick:			
30	Eastern New Brunswick	132	77	Moncton (rail centre). Oysters.
31	Southern New Brunswick	150	77	Saint John (year-round port, industry).
32	Saint John Valley	157	69	Fredericton (capital). Potatoes. Wood. Links with Quebec and Maine.
33	Northeastern New Brunswick	157	61	Minimum index. Wood. Belledune industrial complex. Acadian influence.
	Quebec:			
40	North Shore and New Quebec	81	88	Trade in primary products (iron, wood, hydro power). Linear and scattered ecumene. Ports. Marked gradient of nordicity.

TABLE 9.3 (Cont'd)

Code	*Major Regions*	*Total Population, 1961 (thousands)*	*Market Index*	*Leading Economic Features*
41	South Shore and Gaspé	401	46	Minimum index. Declining primary economy. Weak regional organization. Development plan.
42	Saguenay and Lake Saint John	262	73	Hydro power and aluminum. Wood. Intercity rivalry.
43	Metropolitan Quebec	644	72	Quebec (capital; service centre). St. Lawrence ports. Shipping industry.
44	Mauricie	301	73	Various industries. St. Lawrence Bridge. Trois-Rivières. Shawinigan.
45	Eastern Townships	462	71	Textiles and asbestos. Sherbrooke.
46	Montreal fringe	737	77	Varied economy: agriculture (tobacco), manufacturing, tourism (Laurentides). Dominated by Montreal.
47	Metropolitan Montreal	2,019	114	Port. Industries. Services. National importance.
48	Western Laurentides	182	83	North Shore of Ottawa Valley. Hull. Influence of Ottawa.
49	Western Quebec	169	76	Temiscaming. Abitibi. Mining. Rouyn-Noranda. Links with Ontario.
	Ontario:			
50	Eastern Ontario	782	108	Ottawa (capital); government services. Mixed agriculture.
51	Southern Ontario (north of Lake Ontario)	335	95	Agriculture and tourism. Peterborough.
52	Metropolitan Toronto	2,087	129	Toronto (capital). Large and diversified economy. From Oshawa to Oakville.
53	Niagara Peninsula	762	126	Hamilton (steel). Hydro power. Orchards. Welland Canal. Nearness to U.S.A.
54	London and Erie	405	109	Focused on London. Specialized agriculture.
55	Lake St. Clair	450	116	Heavy industry. Vegetables. Influence of U.S.A. Windsor. Sarnia (pipeline).
56	Midlands	373	107	Kitchener. Mixed agriculture. Engineering.
57	Georgian Bay	319	83	Minimum index. Mixed agriculture. Summer tourism.

TABLE 9.3 (Cont'd)

Code	*Major Regions*	*Total Population, 1961 (thousands)*	*Market Index*	*Leading Economic Features*
58	Northeastern Ontario	506	116	Nickel, copper. Steel (Sault Ste. Marie). Paper. Hydro power (drainage to James Bay). Sudbury.
59	Northwestern Ontario	216	112	Iron. Paper. Flour milling. Breaks in the settlement pattern. Thunder Bay. Links with the Prairies. North of southern Canada: Middle North.
	Manitoba:			
60	Metropolitan Winnipeg	476	112	Winnipeg (capital). Service and manufacturing centre. Former dominance over the whole prairie region.
61	Southeastern Manitoba	80	84	Agriculture.
62	Interlake	40	71	Marginal agriculture. Development plan.
63	Central Manitoba	67	68	Mixed farming. West of the Red River. Portage-la-Prairie.
64	Southwestern Manitoba	153	80	Grain growing. Brandon.
65	Dauphin	58	66	Minimum index. Marginal agriculture.
66	Northern Manitoba	47	122	Railway. Mines (copper, zinc, nickel). Lynn Lake. Churchill. Thompson.
	Saskatchewan:			
70	Southeastern Saskatchewan	227	92	Commercial agriculture (wheat). Regina (capital).
71	Southwestern Saskatchewan	149	73	Wheat and rearing. Moose Jaw and Swift Current.
72	Saskatoon Plain	187	89	Wheat and agricultural products.
73	North and east of Regina	129	63	Agriculture. Potassium. Influence of Manitoba.
74	Central Saskatchewan	212	62	Minimum index. Agriculture. Prince Albert.
75	Northern Saskatchewan	21	90	Uranium mining. Influence of Alberta.
	Alberta:			
80	Southeastern Alberta	39	95	Irrigated horticulture. Medicine Hat.
81	Lethbridge Region	114	93	Grain. Irrigation. Rearing.
82	Rockies and foothills	20	116	Tourism (Banff and Jasper).
83	Calgary region	371	116	Agriculture. Fuels. Manufacturing. Services. Gas pipelines.

TABLE 9.3 (Cont'd)

Code	Major Regions	Total Population, 1961 (thousands)	Market Index	Leading Economic Features
84	Red Deer	76	91	Mixed agriculture. Rearing.
85	East of Edmonton	111	84	Mixed agriculture. Lloydminster. Influence of Saskatchewan.
86	Metropolitan Edmonton	456	112	Capital. Agriculture. Fuels. Manufacturing. Services. Pipeline.
87	Northeastern Alberta	47	83	Marginal economy. Rail and water transport (Athabasca).
88	Northwestern Alberta	96	81	Minimum index. Primary activities. Grain in the Peace River. Railway to Great Slave Lake.
	British Columbia:			
90	Eastern Kootenay	34	138	Maximum index. Mining. Forestry. Influence of Calgary.
91	Western Kootenay	71	134	Refining and paper making. Dams. Trail and Castlegar.
92	Okanagan Valley	95	96	Minimum index. Irrigated agriculture. Rearing. Tourism. Kelowna on the lake.
93	Interior of southern British Columbia	66	115	Agriculture. Kamloops.
94	Metropolitan Vancouver	907	117	Metropolis. Manufacturing (forest products). Agriculture of Fraser Valley. Rail terminus. Port. Services. New Westminster.
95	Vancouver Island	291	118	Wood products. Fisheries. Victoria (capital). Nanaimo.
96	Northwestern British Columbia	59	132	Mountains. Forests. Fisheries. Aluminum (Kitimat). Rail terminus (Prince Rupert).
97	Interior of central British Columbia	74	114	Rearing and wood (paper making). Prince George (focus of communications).
98	Northeastern British Columbia	31	108	Alaska Highway. Fuels (pipelines, refining). Agriculture. Links with Alberta.

SOURCE: Based on P. Camu, E. P. Weeks and Z. W. Sametz, *Economic Geography of Canada.*

Although in the megaprovinces of southern Canada the geographical boundaries deviate little from administrative ones, this is not the case in the North. The Newfoundland portion of Labrador is closely associated with Quebec; "the Labrador-Quebec boundary is of course an absurdity."[9] In Franklin District, there is no similarity between the various types of boundary. The 60th parallel has very little meaning.

Conclusion

In Canada, more than elsewhere, regional disparities pose grave problems, and go so far as to raise questions concerning the structure of the nation. In the confederation formula, the constitution recognizes two main levels of power, the federal government and ten provincial governments. Each of these possesses either concurrent or sovereign jurisdiction in regard to taxation, a part of the income from which may reasonably go towards regional economic improvement. But these sources of finance yield very unequal amounts of money. More than the rich provinces like Ontario, the poor provinces have urgent need of the money raised by the central government. The Atlantic regions cannot raise from within their borders sufficient funds to provide adequate economic and social investments required for redevelopment and action against pollution. These provinces turn, therefore, to the central government for additional assistance. During the 1929 Depression, the Prairie Provinces severely criticized the central government for not doing enough for them. Since its entry into Confederation (1949), Newfoundland has benefited greatly from federal government assistance. Such needs and benefits obviously affect the opposition that the provincial governments might otherwise provide to the extension of power by the central government. However, for cultural reasons, Quebec has always been wary of dominance by Ottawa, and dislikes all administrative formulae which might support federal superiority (medicare; family allowances). Hence, the methods of financing become an element in regional division and even in the definition of Canada.

---

9. T. Lloyd, "Trends", in J. Warkentin, ed., *Canada. A Geographical Interpretation*, p. 584.

Part Three

Types of Canadians

In terms of population numbers, Canada is no longer an insignificant country. There are more than 22 million inhabitants; this was the size of the population of Victorian England and of the United States less than a century ago. In the area around the Windsor-Lévis axis, where 60 per cent of Canadians live, the population density is not much below that of France. The Canadian population possesses three fundamental characteristics: it is mobile, it is ethnically mixed and it is urbanized (see part five). Although the influence of natural conditions is not forgotten, this part emphasizes the historical origins, the regional differences and some geopolitical problems of the population.

Chapter 10

The Growth of the Population

Throughout this study of growth, the same territorial unit will be used, that of present-day Canada. This means that for the colonial periods of French colonization, the French and English populations of Acadia are added to those of the St. Lawrence, and that, up to 1949, the population of Newfoundland is also added. Some additions for certain ethnic groups must also be made to the figures. The native peoples came very belatedly into the official censuses; in view of the active or passive roles that they played during the colonial period (they represented about 60 per cent of the total population in 1763), they should be included in a geography of the Canadian population. There were more than 200,000 indigenous inhabitants at the time of European discovery, but only 100,000 by the beginning of the twentieth century. In the intervening decades, successive estimates show a concave curve. Thus, around 1750 the spatial and ethnic corrections described above increase the "Canadian" population from about 55,000 to more than 240,000. Population growth has also been extrapolated in table 10.1 for the next decennial census (1981). Despite its weaknesses, we thus have a table of population change over four centuries.

Canada since 1600

The first feature of the table is the very concave trend of the overall growth. The total number of inhabitants living within the present limits of Canada has not changed uniformly; population has only grown rapidly in the nineteenth and twentieth centuries. In general, four distinct phases may be recognized.

(1) In the Seventeenth Century

During this period, there was a slow decline in numbers.

TABLE 10.1 Total Population of Canada, 1601–1981

Year	*1600*	*1700*	*1800*	*1900*
01	200,000	199,600	535,000	5,605,315
11	198,067	203,500	696,000	7,464,843
21	196,229	212,200	942,000	9,058,949
31	194,601	225,000	1,291,500	10,660,286
41	193,160	237,000	1,873,000	11,810,655
51	192,430	251,000	2,581,297	14,009,429
61	192,350	265,000	3,405,133	18,238,247
71	195,000	277,100	3,850,557	21,568,315
81	196,000	320,300	4,517,810	25,500,000
91	196,490	404,500	5,046,239	

NOTE: From 1601 to 1781, more than half the population is estimated. Between 1791 and 1871, the proportion which must be estimated declines from 38 per cent to 2 per cent. From 1881 to 1941, estimates account for less than 1 per cent of each total. The totals for 1951 to 1971 contain no estimates. The total shown for 1981 is the author's forecast.

(2) *From 1711 to around 1775*

This second period was marked by a slow increase. The total population recovered and passed the initial level of 200,000 people. The maximum decennial rate of increase reached 6 per cent; at the end of the period, native peoples still accounted for half the total.

(3) *From 1775 to 1861*

In less than a century, the total population exploded from less than 300,000 to more than 3 million; the decennial percentage rate of increase generally exceeded 30 per cent and reached 45 per cent between 1831 and 1841. During the final decade of this period, the population increased by an unprecedented total of 800,000 individuals. Never again in the history of Canada will there be such a large relative increase, certainly not over a period as long as this one (1796–1861). Demographically, this change seems more important than that of the settlement of the Prairie Provinces at the turn of the century. The chief events with which it was associated were the arrival of the British, the settlement of the United Empire Loyalists as a result of the American Revolution, Irish immigration provoked by the economic crisis, the need for manpower after the initial opening up of the country, immigration and the natural increase of the existing population, notably the French Canadians—all of these contributed to the very great increase of population. In relation to the ecumene, the new inhabitants consolidated and created new scattered nodes of settlement in the Maritimes and in Quebec; above all, they opened up southern Ontario. In short, it was eastern Canada which benefited from this increase.

This was the period of the British supremacy over the Indian and even the French strains. Though, in terms of exploration and discovery, the British colonies were almost contemporaneous with New France, they were much slower in the settlement and growth of population; even in Quebec, the first English population revival was qualitative rather than quantitative, represented by influential businessmen such as Molson, McGill and McTavish. Finally, from the beginning of the nineteenth century and particularly towards the end of this period, there was an appreciable emigration of Canadians; at the time of the 1860 U.S. census, 225,000 Canadians were living in the neighbouring republic. This was a foretaste of the great mobility of the population and the permanence of migrations of all kinds.

(4) *The Last Century*

Up to this stage, one major tendency had characterized each of the population phases: a slight decline in the seventeenth century, modest growth during the first three-quarters of the eighteenth century, spectacular increase in the first half of the nineteenth. The period since 1861 is the most important in the history of population; continuous growth took the population from 3 million at about the time of Confederation to 5 million at the turn of the century, then to 10 million during the depression and to more than 22 million now. In contrast to the sustained vigour of the years 1780–1850, this phase has been one marked by four distinct stages.[1]

(i) *1861–1896*. The decennial rate of growth, which had exceeded 27 per cent for the previous eighty years, fell to 10 per cent between 1861 and the end of the century. There were various reasons for this change. Immigrants became less numerous after 1857, following a period of high immigration. The massive emigration which had first affected Quebec during the preceding period was noticeable in Ontario by 1861 and in the Maritimes twenty years later. From 1851 to 1896, for every three new immigrants, two persons left the country. The rate of natural increase also began to decline. None of these features is surprising in view of the world economic crises and the slow increase in the agricultural acreage. The political effects of Confederation and the transcontinental railway do not seem to have been able to prevent some stagnation.

(ii) *1896–1914*. The revival of activity was sudden and the decennial rate showed an increase of 22 per cent during the single decade 1901–1911. On its population map, Canada saw the effect of the western extension of the ecumene: the change in pace was tremendous. The population, which had not quite been able to double itself between 1851 and 1891, now did so in twenty years. During the single decade 1901–1911, the total population increased by almost 1,900,000 individuals. The ethnic composition also changed: native peoples, then at their lowest point, were almost insignificant (1.2 per cent); the French-speaking component declined slightly to 27 per cent (1921). Though the British retained a comfortable superiority in numbers, their proportion fell as the proportion of New Canadians of diverse origins, particularly Germans, rose. Demographically and sociologically, two white ethnic groups no longer clearly defined Canada. World War I affected the length of this period of rapid growth.

(iii) *1914–1942*. The third, very varied stage bears a strange resemblance to the first, through its duration (three decades), its general character (low rates of change) and its recorded maximum and minimum growth rates. Despite the beneficial effect on output of the war and despite the brief postwar prosperity and progressive urbanization, the rate of growth of the population decreased. The major decline in immigration between the wars, the lack of new, important agricultural horizons, the decrease in the birthrate, and more especially the long and terrible decade of economic crisis, checked more positive tendencies. The decrease in the

1. T. R. Weir, "Population Changes in Canada, 1867–1967", *The Canadian Geographer* XI (1967): 197–215.

birthrate was to lead to future shortages in the number of adults.

(iv) *Since 1942.* Again an economic recovery took place and the rate of population growth climbed from 10 per cent to 30 per cent. Wartime activity, a strong and sustained birthrate, massive immigration recalling that at the start of the century, and the growth of certain sectors of postwar production account for the revival in this period. Urbanization emphasized the dominance of eastern Canada, which the opening up of the Prairies had tended to diminish. Canada became a nation of cities. The population, always mobile, became involved in great migrations within Canada: a high percentage of townspeople changed their home between each census.

These periodic variations in the pace of growth emphasize the fragility of the economy and the fact that it is influenced by external events as much as by internal conditions. A broad perspective might suggest that the demographic evolution of Canada takes place in phases lasting twenty to thirty years. If this pattern is sustained, the changeover may already have happened, due to the decline in the birthrate in the last ten years, the absence of massive immigration and the economic problems of the early 1970s.

The Megaprovinces

The overall demographic evolution of Canada hides the very different patterns displayed by its regional components. All provinces have not been affected at the same time and in the same way (fig. 10.1).

From the outset, the major regions supported unequal populations because the West alone contained half the Indian population of the country, while the Appalachian colonies had only a few thousand native peoples. Ontario had twice as many Indians as Quebec. The two emptiest regions were those first affected by European initiative.

Changes in nonnative population emphasize the successive roles played by different regions. In the first period, that is to say, before the nineteenth century, Quebec and the Maritimes contained almost the entire white population. From 1841, only half a century after the arrival of the Loyalists, the population of Ontario exceeded that of the four Atlantic Provinces and, only ten years later, it became larger even than that of Quebec. It was the second Canada, in terms of demographic history. Belatedly, the third period added western Canada to the preceding contributions and for half a century growth was very rapid. Around 1905, the total population of the Prairie Provinces exceeded that of the Atlantic Provinces, and, about 1920, it was for a brief period greater than that of Quebec (the population of the Prairies went from about a quarter of a million in 1891 to 3 million in 1931). Within these megaregions the provincial units have evolved in different ways.

Atlantic Provinces

Despite its name, Newfoundland is Great Britain's oldest colony, and both fishing and settlement began early. At the beginning of the eighteenth century, Acadia—English and French at the same time—and then New Brunswick each overtook Newfoundland. The Maritimes, which had a pattern of development similar to that of Quebec from the seventeenth century to the first quarter of the nineteenth century, have become slower to develop than the rest of southern Canada. Like New England, they have suffered from being outside the dynamic east-centre of North America. During the third quarter of the nineteenth century, the Maritimes, which had already been affected by a decline in immigration, suffered fairly severe emigration. After the end of the nineteenth century, little Prince

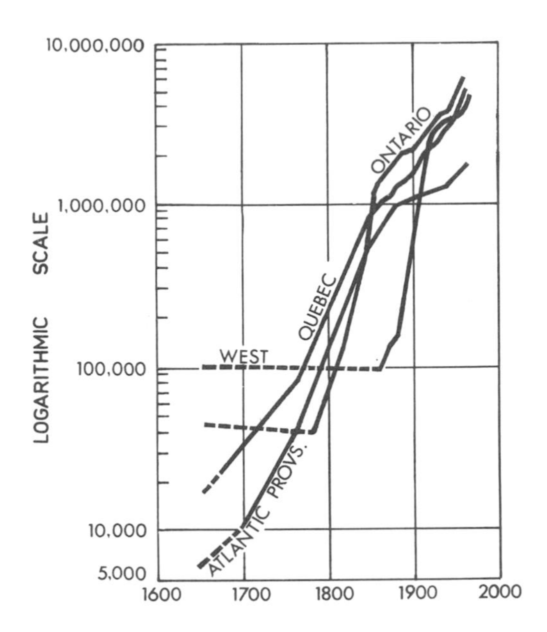

Fig. 10.1 Demographic increase by megaprovince, 1650–1971.

Edward Island, which until then had been growing at almost the same rate as the other three provinces, became almost stationary (population 109,000 in 1891, 111,000 in 1971). In general, despite social and economic differences among the four Atlantic Provinces, they are working towards a similar future.

Quebec

From the time of its discovery, Quebec, with few indigenous people, had a lower total population than Ontario, and particularly than the West. However, it was in Quebec that white settlement came soonest and quickest. From the beginning of the eighteenth century, the French and Indian population of Quebec exceeded the population born in Ontario. Half a century later, Quebec had the largest population among the megaprovinces, displacing western Canada. Towards the middle of the nineteenth century, the descendants of the Loyalists, the large British immigration and the emigration of French Canadians made Ontario pre-eminent. Since 1910, Quebec and the Prairies have competed for second place.

Ontario

The pattern of past population growth in Ontario has three major phases:

(1) From 1770 to 1850, the rate of growth was one of the most rapid in Canadian history, with an increase of a million people.

(2) In contrast, when agriculture in Ontario reached its limits and the Prairies were opened, the decennial rate of growth slowed. For a half-century after 1861, the percentage rate of population growth in Ontario was sometimes the same but more frequently below that of either Canada as a whole or Quebec. In the 1891–1901 decade, the rate of growth fell as low as 3 per cent.

(3) Since World War I, urbanization has been responsible for a third phase of growth. Between 1866 and 1971, Ontario alone accounted for almost half the growth of population in Canada.

Western Canada

Western Canada also has distinctive regional patterns of growth, from the period of discovery when, thanks to the Indians, it had the highest population in Canada. The two outer provinces contained the oldest white settlement. Around 1825, Manitoba had about 2,000 nonindigenous inhabitants; thirty years later, British Columbia had 3,000. The other two provinces, Saskatchewan and Alberta, only appear as such in the first twentieth-century censuses. Apart from the recent character of this white settlement, the West from 1880 to 1920 was characterized by rapid population growth similar to that of Ontario at the beginning of the nineteenth century.

For the last fifty years, each of the four western provinces has had a distinctive pattern of growth. Manitoba, the oldest province, has been exceeded in turn by Saskatchewan (about 1910), then by Alberta and finally by the more stable British Columbia. Saskatchewan eventually reached a ceiling more severe than that of Manitoba. After having been the largest western province in 1931, it has now become the smallest. The result was a great migration; since Saskatchewan had specialized very much in wheat and was urbanized only to a small extent, it had little economic strength to resist the depression and drought and to protect itself from depopulation towards its more attractive neighbours. Potash and fuel industries have helped to restore the situation to some extent.

Even Alberta was affected by a slowing down in growth about 1930, but oil discoveries and a more diversified economy allowed it to recover to the point where, from 1951 to 1961, its rate of growth was the fastest of any province. British Columbia's population growth, on the other hand, has been much more gradual. Apart from a few unusual periods, such as the Cariboo gold rush, which had the beneficial result of opening a way into the interior, economic growth has been relatively continuous. Growing like the Pacific states of the U.S.A., British Columbia now has the largest population of any of the western provinces (2,184,671 in 1971), exceeding the combined populations of the four Atlantic Provinces.

At the 1971 census, the regional distribution of population was: Territories, 0.2 per cent; Atlantic Provinces, 9.5 per cent; western Canada, 26.6 per cent; Quebec, 28.0 per cent; Ontario, 35.7 per cent. Two of the units thus contributed almost equal proportions of the Canadian population. Ontario is moving ahead.

The Canadian population is prolific and mobile. The birthrate, internal migration and immigration, the death rate and emigration have all played significant roles in the development of the population. The present situation has deep roots, and the past must again be taken into account.

Chapter 11

The Origin and Development of Canadian Demographic Characteristics

Biological Aspects

For a country with a high standard of living, the Canadian population is biologically active. During the last hundred years, the decennial rate of natural increase has always been higher than 14 per cent (except in the 1931–1941 decade) and it has often reached 18 or 19 per cent; this represents an annual rate of total population increase which averages 1.5 to 2 per cent. Canada was sixth in the world in terms of rate of increase in 1961, behind some Latin American countries which have very high birthrates and moderate death rates. In 1966, life expectancy at birth reached seventy-five for girls, a longevity which is a tribute to medical and public health programs.

Changes in the Birthrate

In North America, the level of economic activity markedly and rapidly influences individual behaviour. The gross birthrate, which had fallen during the depression of the second half of the nineteenth century, remained high during the years of large-scale immigration at the turn of the century. In 1921, at a time of prosperity immediately after World War I, the rate reached 30 per thousand, but in 1937, during the depression, it fell to 20.1 per thousand, the lowest rate recorded up to that time. World War II, and the postwar period even more, raised it again to 29 per thousand. The gross reproduction rate (the average number of girls born for each woman aged fifteen to forty-nine years) reached almost twice the level required to assure replacement. The recovery after 1942 depended on several factors, including new attitudes to the family, a large number of immigrants and a general mood of optimism. This recent demographic revival is perhaps the most important event in Canada since the opening of the West at the beginning of the century. During the 1951–60 decade, births were practically double those of the 1931–1940 period; the marriage rate similarly increased from 6.5 per thousand between 1931 and 1935 to nearly 10 per thousand in the years 1946–1950. A fertile generation is therefore following one which was much less productive. Around 1960, this sequence made the proportion of adults relatively low in comparison to that of young people. This age structure contributes to the relative power possessed by Canadian youth, a situation which will no longer exist in the 1980s.

Canada is now experiencing a marked fall in the total number of births, however, a trend which began in 1959 and increased after 1963. Despite short-term changes, the secular tendency towards a decline in the birthrate has become evident again. The rate decreased from 45 per thousand (1851–1861) to 34 per thousand (1881–1891), then to 21 per thousand (1936–1940), reaching 17 per thousand in 1971. Never, even during the depression, has the rate been so low; it is without doubt due to family planning and the use of contraceptives, to urbanization, to working wives and to improvements in the standard of living.

What has this meant for Quebec's "revenge of the cradle"? Traditionally, there has been a contrast between the fertility of French Canadians (the gross rate reaching a maximum of 65 per thousand around 1765, according to J. Henripin)[1] and the English-Canadian "Malthusianism", expressed as a low birthrate. This contrast has now disappeared; for the last thirty years, several English-speaking provinces have, at different times, recorded gross birthrates higher than that of Quebec. At the 1961 census, five of them, the Atlantic Provinces and Alberta, had higher rates. In the same year, a study of the gross fertility rate of women of childbearing age showed in fact that Quebec (122 children per thousand women) was next to the lowest in the list of provinces. The decline of the birthrate particularly affects the size of the completed family. The pattern in Quebec, which has become more sensitive to socio-economic conditions, is therefore no longer the traditional one. Quebec family allowances do not seem to have changed the trend.

Deaths

The number of deaths naturally varies less than the number of births, excluding the world wars, which directly or indirectly accounted for more than 150,000 deaths, and excluding also the death rate of the native peoples, which rose until the twentieth century. The gross death rate is diminishing; it reached a maximum of 38 per thousand in Quebec during the decade of the French defeat (1751–1760). For Canada,

1. See J. Henripin, *Trends and Factors of Fertility in Canada*, 1961 Census Monograph (Ottawa, 1972), p. 5.

it was over 20 per thousand between 1861 and 1871; since 1900, after remaining for a long time in the neighbourhood of 13 per thousand and then about 10 per thousand from 1931 to 1940, it has fallen to less than 8 per thousand and even to below 7 per thousand in some provinces. In 1968, the Yukon had a rate lower than 6. As the average age of death has risen, the illnesses of old age are more common than they used to be, whence the increased frequency of cancer and cardiovascular diseases. In 1968, Canada had the third-lowest death rate in the world, and may be compared to the United States (fifteenth from the bottom), France (twenty-fourth) and Great Britain (twenty-sixth).

Overall Natural Increase

Under such conditions, the balance of births minus deaths must be overwhelmingly positive. During the century which preceded the 1961 census, there were over 22 million births as compared to 10 million deaths. Each of the ten decades provided a surplus, and since 1900, natural increase has exceeded the number of deaths. This is the major reason for the general growth in the Canadian population; for the century up to 1961, the total number of immigrants was 8 million, while natural increase totalled more than 12 million. Since, as we shall see, the majority of immigrants soon left again, the real contrast between the contributions of natural increase and immigration is very great; internal growth dominates external additions. From 1951 to 1961, 72 per cent of the total increase in the Canadian population was due to natural increase; the remainder came from immigration. This dominance of internal growth has characterized eight of the eleven last intercensal periods.

During the first two decades of the twentieth century, the number of temporary immigrants slightly exceeded the natural increase. It was specifically during the years 1903–1913 that the balance was significantly reversed: 2,600,000 immigrants arrived, while natural increase represented only half this number. A better equilibrium was soon achieved, since the war put a curb on new arrivals, emigration removed a large number of the new immigrants and births to Canadian residents were relatively high before 1931.

In a developing country, it is significant that the growth of total population should depend mainly on natural increase; ever since the first century of the colonial period, those born in the country have been more numerous than those arriving from Europe. Without this dominance of Canadian-born in the population, the creation of a national identity would scarcely have been possible.

Population Movements

Two types of movement must be examined, that across the external boundaries of Canada, and internal migration. During the decade 1911–1921, immigrants, emigrants and interprovincial migrants represented almost 40 per cent of the total Canadian population. Thus, population movements are a large-scale phenomenon.

International Migration Balance

The importance of these movements is well known. Before the twentieth century, the non-native population of Canada came almost exclusively from French and British stock. Subsequently, millions of immigrants from a variety of origins took part in the opening up of the Prairies and in the growth of cities.[2] Equally well known is the emigration of many of these new arrivals and of French Canadians

2. See W. E. Kalbach, *The Impact of Immigration on Canada's Population*, 1961 Census Monograph (Ottawa, 1970).

to the United States. During the last hundred years, Canada gained 8 million people and lost 6 million; at least 14 million individuals crossed her international boundaries, a total equivalent to the population of the whole country in 1951. During the decade 1901–1911 alone, immigrants and emigrants represented one-third of the resident population. Such movements have therefore been huge and, in relation to the resident group, disturbing. It is astonishing that Canada, so far from the countries where emigrants originate, has received so many immigrants. In 1972, Canada welcomed its ten millionth newcomer. It is even more startling to realize the great number that leave the country. Who are the ones who have gone? Do people from other countries arrive to take the places of those who were born here, or are immigrants to Canada really only nomads?

For the century before 1961, for four people who came in, three would leave; among the latter, immigrants accounted for more than 4,500,000. If these calculations are reasonably accurate, they suggest that only about 40 per cent of those who have come from abroad remain in Canada. The proportion seems higher now and from 1966 to 1970 it increased to 60 per cent. Immigration has nevertheless been an essential factor in the history of the country.[3]

If, over the century as a whole, those who have come in have exceeded those who have left, this surplus has not been a permanent feature. There have in fact been two sets of contrasting conditions. After several decades with a favourable balance, the period 1861–1901 showed a negative total; not only did as many people leave Canada as come in, but almost 500,000 more went with them. The second period (apart from a slight reversal of the situation during the depression of the thirties) has lasted from 1901 and is marked by a positive balance, especially from 1951 to 1961. It seems probable that the balance of migration will become less and less disadvantageous. For example, take the balance of migration between the U.S.A. and Canada from 1954 to 1964. During this period, 320,000 Canadians went south, while American immigrants to Canada numbered about 100,000 and Canadians returning from the United States numbered at least 50,000, therefore leaving a deficit of around 170,000. After deducting from this residual number those who have since died and those who have the firm intention of returning to Canada, the deficit is really insignificant in comparison with a Canadian population of more than 20 million. Immigration is becoming less significant; in 1971 the total number of immigrants represented only 0.5 per cent of the Canadian population.

Population Movements Within Canada

The psychological and physical mobility of the Canadian population is not merely an international affair, for it also occurs intranationally, and involves both immigrants and residents. In spatial terms, movements are from one province to another, or, more difficult to trace in the statistics, from one part of a province to another. To a very great extent, it is these intra-Canadian migrations that have led to urban growth.[4]

Over the last hundred years, 5 million Canadians have moved from one colony or from one province to another. The size of the total must be emphasized, because it is probably twice the number of nonimmigrant Canadians who have emigrated abroad and it is not far from the total external emigration (native-born and foreign-born Canadians). The population of Canada is indeed an extremely mobile one.

3. See [Canada], *Atlas of Canada*, 1957 edition, plate no. 46.

4. R. Lycan, "Interprovincial Migration in Canada: The Role of Spatial and Economic Factors", *The Canadian Geographer* XIII (1969): 237–254.

Intraprovincial migrations are also found throughout the country and each of the provinces, even Quebec, which may seem more stable, has always both provided migrants and received them from elsewhere in Canada. Over the long term, migrations from one province to another have become increasingly important. At the time of Confederation, they affected only 2 per cent of people born in Canada, but the figure reached 6 per cent at the beginning of the twentieth century, thanks to the railways and to the settlement of the Prairies. By the 1941 census, the proportion was 11 per cent. In recent decades, such movements have involved a greater number of people than either immigration or emigration, as Canada tends to redistribute its national population (see table 11.1).[5]

The various provinces have not all been equally able to hold or to attract population. Ever since 1871, the Maritimes, with a substantial net loss of people, have contrasted with British Columbia's positive balance; other trends, however, require the introduction of a time frame. Manitoba, until 1901, and Alberta and Saskatchewan, until 1921, gained substantially from the eastern provinces; interprovincial migrations provided something like half the initial population of the Prairies, mainly from 1881 to 1901. The period after World War I had a very different effect on the industrial provinces (Ontario and Quebec) than it had on the agricultural Prairies. As urbanization has quickened, it is the urban provinces of Ontario, Quebec and British Columbia that have absorbed the large majority of interprovincial migrations.

About 1960, these movements tended to follow standards of living; there has been a transfer of people from provinces with low economic levels to provinces with high levels as the poorer regions of the Maritimes, Quebec, Saskatchewan and Manitoba have lost inhabitants to the cities of British Columbia, Alberta, Ontario and Quebec (Montreal). Note that the evolution of Alberta tends to be different from that of the other Prairie Provinces; also, Quebec differs from Ontario.

The Canadian population is, therefore, like a patch of sand blown by the wind in a variety of ways; individual elements are picked up and dropped again. The shallow-rooted elements move from one place to another in search of success. How far they move varies; the majority remain within the same province. Some links are between adjacent provinces, as from Saskatchewan into Alberta; others are long-distance, as from Newfoundland to Ontario; and some are across Canada, from the Maritimes to British Columbia. A population which is nomadic rather than transhumant is a realistic basis for the old French-Canadian song about the *Canadien errant*. Canada provides a good example of the increasing mobility of the world's population.

Reasons for and Regions of Migration

The reasons for migration, whether external or internal, depend on the period and the cultural group concerned. Some factors are very common, like the lack of deep roots in many families. This is translated into a flexible attitude to movement: one's country is not one's place of birth, but where one can make the best living. Without the individual's being fully conscious of it, this attitude reflects a refusal to be constrained by the limitations of one's surroundings and emphasizes instead a personal desire to be successful. This drive to succeed, which involves the willingness to adopt a life style perhaps requiring several moves in one's time, is at the heart of the capitalist spirit. Such mobility is not limited to immigrants; it affects even the allegedly sedentary peasant.

5. See M. V. George, *Internal Migration in Canada. Demographic Analyses*, 1961 Census Monograph (Ottawa, 1970).

TABLE 11.1 Net Migration Per Thousand Canadian-Born, by Province, 1871–1961

Period	*Nfld.*	*P.E.I.*	*N.S.*	*N.B.*	*Que.*	*Ont.*	*Man.*	*Sask.*	*Alta.*	*B.C.*
1871–1881	*	*	−5	4	−9	−11	*	*		*
1881–1891	*	−18	−6	−11	−8	−28	283	355		371
1891–1901	*	−31	−5	−8	−3	−29	128	244		242
1901–1911	*	−59	−29	−26	−14	−76	−71	743		308
1911–1921	*	−34	−13	−1	−7	−1	−57	23	86	75
1921–1931	*	−9	−20	−17	10	21	−85	−79	−37	95
1931–1941	*	−15	−10	−17	9	35	−68	−176	−62	116
1941–1951	*	−60	−48	−62	−3	45	−100	−230	−15	197
1951–1956	*	*	*	*	*	*	*	*	*	*
1956–1961	−9	−8	−20	−6	−1	73	−12	−32	17	39

▭ area of positive net migration

SOURCE: M. C. Urquhart and K. A. H. Buckley, eds., *Historical Statistics of Canada*, and 1961 Census.

NOTE: Asterisks indicate that data are not available.

Surveys suggest that less than 10 per cent of land-based families in Quebec have lived on the same farm for more than three generations. Many people are, therefore, only "conditional" residents; if they are not satisfied they move, because they are mentally adjusted to moving.

This tendency to move is important in regard to four groups of regions, three of which are foreign countries. First, there is the United States, which has had to absorb the majority of emigrants (more than 60 per cent). As Siegfried has written, Canada is attracted towards her large neighbour in a manner which is "indefinable, persistent and, to be truthful, fatal";[6] the long ribbon of ecumene in southern Canada assists the development of this attraction. The bond of language between the powerful United States and more than 70 per cent of the Canadian population creates an empathy in which the position of Canada becomes that of a satellite. In the neighbouring republic, economic activity and employment opportunities are greater, salaries are higher, winter less severe. With all this, the attraction has been, and often still is, irresistible, particularly when it calls to an immigrant for whom Canada is only a springboard to the United States. However, in recent years, the social problems in the States have restrained many Canadians from crossing the border to settle.

The British Isles is an area where the attraction is primarily sentimental. People leave the United Kingdom to work in different parts of the Commonwealth, like Canada. In turn, according to 1950 data, about 15 to 20 per cent of the immigrants left Canada for other English-speaking countries. Perhaps as many as a quarter of a million citizens born in Great Britain and counted in the 1961 Canadian census left before 1971. Canada is, therefore, a staging post for pilgrims from the homeland of the English language.

There is also the equally natural appeal of other countries. There are many immigrants who have returned, disappointed or enriched, to Italy, Central Europe, France, and so on. Some Hungarians admitted to Canada after the revolution of October 1956 have since returned to their own country.

6. See A. Siegfried, *Le Canada, puissance internationale*, pp. 9–12.

Lastly, the promised land is often another region of Canada. After 1896, Canadians from all over the country set out for the Klondike. Even the people of Ontario were not able to resist the attractions of prairie settlement (see table 11.1). Internal migrations now tend to follow two types of work opportunity: the major enterprises offering employment (such as iron in Quebec-Labrador, the Dew Line, Kitimat, pipeline construction, Elliot Lake, Manicouagan, the Peace River dam, Churchill Falls, the James Bay project) or the big metropolitan centres, such as Toronto.

The population movements, which are such a permanent characteristic of Canada, vouch for the conditional nature of the habitability of the country. The demographic trends show clearly that Canada has severe limitations in its ability to absorb immigrants and even to retain its own people. For more than a century, if the contemporaneous opportunities available to make a living are taken into account, Canada has not been underpopulated; unemployment has been an almost permanent feature. It would be necessary to adopt a new life style, to benefit from extraordinary technological progress or to find a new international role for Canada if it were to absorb a large part of the too-rapid growth in world population. The present economic development of the country seems to warrant a certain amount of self-satisfaction.

Age Pyramids as Evidence of Population Changes

Age pyramids provide the evidence for dynamic changes in population, especially the major features.

In 1931, and still more in 1941, the pyramid resembles a bell, because of a new and serious narrowing at the base. This was due to the depression and, indirectly, to a fall in immigration. There were also relatively few people between thirty and forty years of age in 1931, which in turn emphasizes the short-lived character of the population surge at the end of the nineteenth century. The lack of births in the depression years is still a significant feature of the 1961 pyramid, in which age groups from twenty to forty were smaller than at any time throughout the preceding hundred years. More recently, postwar prosperity and immigration have broadened the base again (in 1961 it resembled that of 1901) and have added to the size of the adult age groups. To some extent, therefore, the 1961 pyramid reflects directly and indirectly events since 1880.

The depleted age groups from fifteen to thirty-five (especially from fifteen to thirty, due to recent immigration) pose a grave problem; no preceding decennial pyramid has shown such small numbers relative to other age groups. This already small age group will be further diminished by deaths, while its social responsibilities will greatly increase, due to the great number of people in subsequent age groups and the growing numbers in the older age groups. Further, the small size of this group entails unfortunate repercussions on future birthrate. Lastly, an expanding economy, despite technological revolutions, requires a work force to sustain it. On the basis of the 1951 census, the author recommended functional immigration as a remedy: in other words, admit 700,000 foreigners aged from fifteen to thirty years and able to respond to particular needs of the economy. If this entry were spread over ten years it would mean an average of 70,000 per year, a small number when compared to what Canada has taken in for the last sixty years. These immigrants, since they are needed, would not be unemployed and thus would probably not be prospective emigrants.

Canada is now moving, however, into quite a different situation. The large number of births between 1942 and 1963 has resulted in a swelling of the young adult age groups, so that there now seems to be a national excess in the labour force and consequently, serious unemployment problems. In terms of population, Canada at the end of this century may be very different from the transitional status of today. In 1960, it was dominated by youth; by 2000, it will once again be a nation of adults, especially in view of the present low birthrate.

These dynamic shifts in demographic character represent one of the fundamental features of the geography of Canada.

Chapter 12

Cultural Groups and the Growth of Canada

Cultural groups are one of the most important aspects of the geography of Canada. In the view of some political scientists, they were fundamental to the Confederation which created the Canada of 1867, a union founded on two peoples, English and French. In our examination of this aspect, we may also consider the increase in cultural variety during the twentieth century. It is a suitable point at which to mention as well the paradox that in the mid-1960s the Indians, the original inhabitants of Canada, had their affairs handled by a minister of immigration! Because this subject has created great interest, a centre of ethnic studies is likely to be organized in Canada.

The Indigenous Groups

Ever since Christopher Columbus mistook America for the Indies, Canada's natives have been called *Indians*; since this word is used in two countries of the same Commonwealth, confusion has resulted, hence the frequent use of the term *Amerindian* to describe the Indians and the Eskimos of Canada. Sociologically, the word *savage* is unacceptable and it has certainly not been appropriate, if indeed it ever was, since these peoples became sedentary and subject to the rule of our laws. *Eskimo*, which means an 'eater of raw flesh', is also an unfortunate term, as is the word *Redskins* (describing those who once completed the colourful palette of the world's inhabitants), which derives from the habit of the former Beothuks of Newfoundland, who painted themselves with ochre. In biological terms, these separate groups that lack felicitous names do not constitute a population as distinctive as is popularly imagined. Even before the sixteenth century there was intermixing between the southern tribes in the present United States and those in the southern parts of Canada, between forest tribes and arctic ones, between the inhabitants of the east and migrants from the west, possibly between Eskimos, Indians and Norsemen, and certainly later between the French, British and Indians. "I am partly Indian, partly police," remarked one Métis from Mackenzie District. Despite these deliberate or chance intermixings, the native peoples comprise several groups which have adapted in different ways to their environment.

Two definitions of a native are possible, one being ethnic and based on physical characteristics, the other legal, in which the status of an Indian is accorded to each individual on a reserve, despite intermixing. Hence, not all legal Indians are full blooded. It should also be noted that in sociological terms, the word *Indian* has a limited meaning which does not include Eskimos.

Historical Geography

Little is known about the prehistory of the Amerindians. Evidence is not very abundant, as a nomadic life does not encourage the development of significant social structures which might find expression, for example, in the construction of important buildings: "The forest was the church of the Montagnais," wrote Jacques Rousseau;[1] and their wooden defences, which might have provided tangible links to their past, could not survive in a climate alternately humid and frosty. However, though the pieces of evidence are scarce, they provide sufficient proof that the natives were still the first occupants of Canada.

In the preceding chapter on the growth of population, some readers may have been surprised by the inclusion of the indigenous population; its historical role has usually been underestimated. The natives took an active part in exploration, in the establishment of permanent settlement, in development, in exploitation, transport and in the colonial rivalries. The exploration routes and then the fur-trade routes were those of the natives and then of the *coureurs de bois*.[2] Since there were no maps, it was the Indians who identified the navigable rivers, the trails, the portages and the campsites. During the colonial period, the network of communications was based on the drainage pattern and on the coasts, both of which were well known to the Indians. The Europeans borrowed not only guides and routes, but the means of movement: they used the fragile bark canoe in summer (they also learned how to repair it rapidly), and they borrowed snowshoes for winter. Those forms of native life that were characterized by a nomadism closely related to wildlife resources were extremely useful to a colonial economy based on the scattered collection of furs. In addition, the existence of broad linguistic groups (extending from the Gulf of St. Lawrence to Saskatchewan) enabled the white man to penetrate a long way into the continent without serious difficulties of communication. Geographically, the collection of fur in Canada depended on the Indian and on Indian methods. This primitive civilization, with its methods of housing, its ideas on clothing, cooking, medicine, language, agriculture (maize and sugar), and even some of its attitudes and outlooks, was made available to the first immigrants. The white man benefited from the prolonged adaptation by the natives to a difficult environment of which the newcomers were entirely ignorant.

Further, and in a wider context, the natives were included among the general aims of colonization. The Christianization of the worshippers of the Algonkian god Manitou was often declared to be one of the colonial objectives.

As a result of the conflicts of economic interest and prestige between France and Great Britain, conflicts which coincided with the territories occupied by rival native groups, the Indians found themselves on the side of one powerful colonizer or the other, although as allies they were not always reliable. Some natives fought beside Montcalm in 1758 during the French victory of Carillon. Later, during the relatively belated settlement of western Canada, Indians and Métis who were already established in the valley of the Red River opposed the advance of white settlement; the famous Mounties had to be formed to combat this resistance, which was finally crushed at Batoche (Saskatchewan). Their leader, Riel, a predomi-

1. M. and J. Rousseau, "Le dualisme religieux des peuplades de la forêt boréale", *Selected Papers of the XXIXth International Congress of Americanists*, Vol. II, 1952, p. 119.
2. See G. L. Nute, *The Voyageur* (St. Paul: Minnesota Historical Society, 1955).

Plate 12.1 An Eskimo cutting open a spotted seal. It is the dog team that indicates this photo was taken as long ago as 1956. By the 1970s most huskies had been retired in favour of snowmobiles.

nantly French Métis, was hanged by a Conservative administration, which caused the great majority of French Canadians to turn against the federal Conservative party.

By the colonial period, the Indian population was already living mainly in the south of Canada, and therefore played an important role during that colonization. The majority of Indians lived in western Canada, as they do at present. The Plains Indians, hunting buffalo, were different from Indians of the east, and came into close cultural contact with the white man at a later date than those in the east. The sedentary groups, fishing salmon on the Pacific and building totem poles, were culturally advanced; the Haida on the Queen Charlotte archipelago had the highest population density in precolonial Canada, about ten times that of the present Yukon. As Diamond Jenness showed in his classic study, the native peoples can be divided regionally into seven categories: the migratory tribes of the forested East (e.g., the Montagnais), the semisedentary groups (Iroquois), the tribes of the Prairies (Blackfoot), Pacific Coast groups (Tlingits), Cordillera mountain peoples (Salish), those of the Yukon and Mackenzie basins (Chipewyan) and, lastly, the Eskimos.[3]

3. See D. Jenness, *Indians of Canada.*

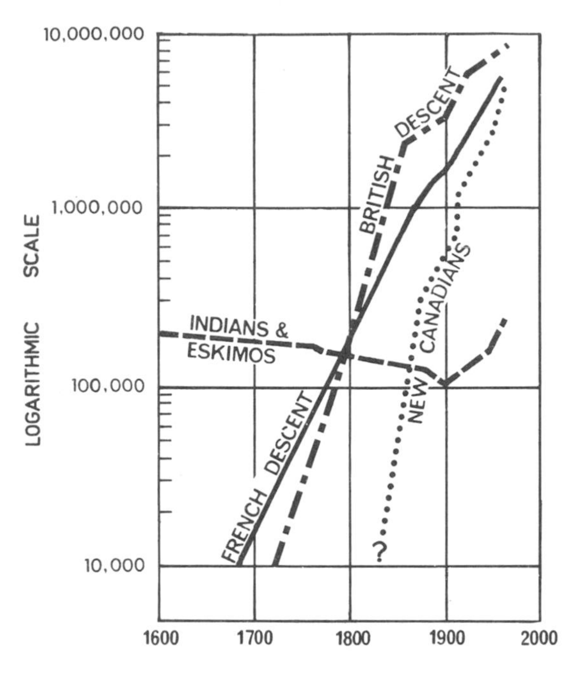

Fig. 12.1 Demographic evolution by ethnic groups, 1600–1971.

For three centuries, the outstanding characteristic of population changes among the indigenous peoples was a slow and gradual decline in numbers (fig. 12.1). From more than 200,000 in the sixteenth century, there were only about 150,000 by 1763; some writers suggest that a minimum of 100,000 was reached about 1900. During this time, the rapid increase in the numbers of other cultural groups reduced the proportion of natives in the total population. From practically 100 per cent in 1660, the proportion was about 60 per cent in 1763; in terms of population, the British were the conquerors more of an Indian territory than of a French one. Then, with British immigration, the population explosion of French Canadians and the arrival of New Canadians, the native proportion fell to 11 per cent in 1831, 3 per cent at Confederation (1867), and 1 per cent today. From being the whole and then the majority, the Indian population has become very much a minority.

This fall in numbers was due to various causes, including the culture shock to the life style of the natives, colonial wars, intertribal wars (e.g., massacre of the Micmacs in 1533 at Bic), infectious diseases brought by the white man (smallpox in the east about 1700; influenza in 1830 on the Pacific coast; typhoid in 1903 on Southampton Island, in the Arctic), the natural or commercial depletion of the caribou, reductions in the prices for pelts, famine, alcohol, an increase in hunting accidents due to the availability of more powerful weapons, and intermarriage between Indian women and white men. If the Europeans had never arrived, the native population would certainly not have diminished as much. Conversely, if it had increased as it has done in the last twenty years (an achievement difficult without the influence of the whites), it would now form the most important population group of Canada. In this respect, Canada is not unlike Latin America.

The tendency towards declining numbers reversed itself at the beginning of the twentieth century. This reversal had begun earlier among the eastern natives and after 1911, the Indian populations of British Columbia and the Prairies also started growing. The Eskimo population, once as high as 22,000, reached a minimum of 7,000 during the first quarter of the twentieth century, and then recovered to the nearly 15,000 which it is today. For the last twenty-five years, the rate of growth has been remarkable, and it has even exceeded the high rates of the nonnative population reached about 1960. This revival is partly due to a fall in the death rate, itself a result of the better services provided by governments. In 1961, the gross rate of natural increase among Eskimos reached 38 per thousand. The population will continue to increase rapidly as, in 1964, 58 per cent of the Indians were less than twenty-one years old, in comparison with 43 per cent of all Canadians. The problem of the demographic survival of the natives, therefore, no longer exists, even though the birthrate among Eskimos has now started to fall. There is, however, less cause for optimism concerning the cultural future of the native peoples.

Acculturation

In legal terms, Indians may have either of two types of status. One was determined by the agreements and historic treaties which defined the rights and duties of tribes and governments. The other status, which is very vague, affects those nomadic bands which did not sign treaties or contracts respecting the lands on which they had previously lived. Today, three-quarters of the Indians live on more than 2,000 scattered *reserves*, areas which are inhabited and are, in some cases, developed (game, fish, wood, agriculture, mining). This gives a nominal density of 16 inhabitants to the square mile (6 persons per square kilometre). The reserve guarantees certain exclusive rights which are generally valued by the Indians. Since the war, population growth, the static character of the natural resources of the reserve, an economic upsurge in the Middle North where some of the indigenous people are found, and a certain cultural awakening have combined to encourage Indians to find work off the reservation. The features of the reserve hence become less significant in employment terms. Economically, many Indians aspire to become citizens like everyone else, despite the need to pay taxes. To assess properly the degree of acculturation of the native peoples, the 560 bands would have to be examined individually; white civilization has affected them in different ways, depending, on the one hand, on the organization, cohesion, resistance to change and capacity to adapt displayed by different groups, and on the other hand, on the opportunities, the timing and the length and intensity of contacts. A distinction must always be made between Indians and Eskimos.

White impact on population numbers is the explanation for both the earlier centuries of decline and the last half-century of increase. White influence has also determined the various occupations. At one time, hunting, fishing and domestic work occupied almost all the native peoples. The change is such that the same index of employment categories used for nonnatives is used for recording the employment of Indians. It is true that there are major differences: there is much more employment for Indians close to nature, but already one-third of Indian workers are in industry and service employment. This tendency will continue. Fishing and hunting occupy only 17 per cent of all workers and many of them are on a part-time basis. Indians can no longer be described, therefore, as hunters or fishermen.

The native labour force has its own characteristics: it is difficult for an Indian to get a job and it naturally takes time for him to adjust to a new working environment. His physical stamina is limited, his basic and technical education is incomplete, and his awareness of these problems may itself be a factor in achieving an equivalent work performance. His wages are low, and the reserve Indian who works at traditional activities earns even less than those who can get work off the reservation. One of the main economic characteristics of Indians is certainly their poverty; most of the families are on welfare for several months of the year. Too many natives are dependent on the state and to too high a degree, a situation which they do not like.

The external conditions of life have also changed. Even among the Eskimos, the tendency to agglomeration is evident, as this is a pattern which is better adapted to the provision of health and education services. Frobisher Bay Eskimos have been described as townsmen. Some cooperatives have been set up to develop

Plate 12.2 Soapstone carving of an Eskimo hunter. The pointed-head parka indicates an eastern Arctic origin.

commercial fishing, retail trade, and handicrafts. Eskimo art is experiencing an unprecedented popularity; the supply of soapstone carvings is not sufficient to satisfy the increasing demand. Pieces from Povungnituk (Quebec) are particularly fine, and this cooperative sold sculptures worth $100,000 in 1965–66. North of the 60° parallel, in 1971, thirty-two Eskimo cooperatives handled more than $5 million worth of business. Since 1963, thanks to government encouragement, a little sealskin owl called Ookpik has become a craze and has even appeared as the official mark of Canada in worldwide fairs. Through agreement with the model's creator, a Chimo Eskimo woman, industrial production has doubled the handicraft output. Sikusi, the ice-worm from the Mackenzie, has been less successful.

The government provides considerable assistance to Indians and Eskimos in health services, basic education (day school and boarding) and specialized training.[4] This training, however, is not always adapted to local needs and is rarely given in the native languages. From every point of view, the provision of schooling has an enormous influence, because it cuts the links between the scholar and his environment: the pupil does not learn to hunt and, when he comes back home, no longer wants to. What happens is not just a break in the pattern of life but the creation of a high social barrier between parents, who are products of their environment, and their uprooted children. Though the generation gap is worldwide, it reaches an extreme when the "Beatle" is the son of an Indian fur trapper. Some would argue that the change has taken place too rapidly.

In terms of mental attitudes, the settler from the south has still not come to understand fully what characterizes the native peoples, beyond an easy smile, silence, carelessness, lack of perseverance and, in short, a passive attitude to life. Socially, the native remains isolated; without being openly subjected to white racial prejudice, he suffers cultural prejudice. The native is a remote figure in a pioneering society which, in contrast to him, is highly enterprising. Half a century after his population low, the native seems to have reached his cultural low. A reawakening of pride is now making itself felt and some native writers are afraid that the system of generous assistance degrades the natives to the level of parasites. A number have begun to occupy responsible positions; as early as 1949, British Columbia elected an Indian member to the legislature. Some groups have their missionaries and their teachers. There is an Eskimo literature using the Roman or syllabic alphabet. Native peoples had their own pavilion at Expo '67. In 1972, four elected members in the Council of the Northwest Territories were Indian or Eskimo. The federal government has created a native cultural council. Eskimo groups want to organize themselves under a regional government. Native peoples who seemed certain to be finally assimilated are now striving to avoid losing their identity. Too much has happened for it to be possible to return to the old days with all their problems; however, life will not be easy for those natives wanting to imitate the development of other groups while remaining themselves. In the years ahead, indigenous peoples could become a national issue in Canada. Indians in the Mackenzie are ready to oppose the building of pipelines to carry Alaskan gas and oil. Conversely, the unemployment rate among native peoples is becoming something that will not be tolerated. In short, Canada will have to give more attention to the culture of the Amerindian.

Canadians of French Stock

There is a superficial similarity between the indigenous peoples and the French-speaking

4. See H. B. Haythorn and M.-A. Tremblay, *Survey of the Contemporary Indians of Canada; Economic, Political, Education Needs and Policies* (Ottawa: Queen's Printer, 1967).

group. They are both components of the population that were here before the massive British influence; they are found in all provinces, they are basically defined by cultural criteria and their economic significance is much smaller than their total numbers. However, the similarities do not go much deeper; in fact, there is an immense gulf between the native peoples and the French Canadians, whose ancestors often played the role of conquerors. Canadians of French origin (5,540,000 in 1961) represent the largest minority group in the population, while the Indians (under 300,000) are one of the smallest. Canadians speaking French exist as an organized society, aided by the powerful political tool of the government of Quebec. Their language is an official one, even in the federal government. Moreover, they have developed powerful religious, cultural, legal and even economic institutions, which are almost nonexistent in the case of the native peoples.

First, the problem of definition. Who are the French Canadians? They are not the French of France, nor the "Français d'ici",[5] even if they are undoubtedly part of the French cultural family. Sociologically, the distinction was established about the seventeenth century and in New France there were even conflicts between the *Canadien*, born in the country, and the Frenchman who had come to work there for a time. The French navigator Bougainville wrote, in 1756: "The Canadians and the French seem to be from different nations." During the Seven Years' War, the "French" army in Canada was divided into regular French soldiers and a Canadian militia. After the French defeat, a decline in personal and cultural contacts with the homeland, North American conditions, and the impact of British and even native influences accentuated the wide gap between French Canadians and the French of France. They are no longer the same people; the majority of Quebec students in Paris have problems of adaptation. To avoid all ambiguity—especially since some French of France are still emigrating to Canada—one cannot describe as French the descendants of that little outpost of Paris and western France which existed on the St. Lawrence. The French-speaking group in Canada includes the French Canadians in Quebec, Ontario and western Canada, the Acadians of the Atlantic Provinces, the French from France who have become Canadian citizens, the French from France who live in Canada but are not citizens, and the immigrants of various nationalities (e.g., Italians, Jews) who have adopted the French language.

The censuses provided two definitions of a French Canadian: one is racial, or ethnic, based on the origin of the father; the other, a cultural definition, is provided by the choice of mother tongue. This distinction is important because there is a numerical difference between French-Canadian stock (French by descent) and the French-Canadian language (Francophone people). Though the difference was small before about 1840, it is becoming larger at present. At the 1941 census, there were 130,000 fewer Canadians who normally spoke French than there were Canadians of French descent. Ten years later, the difference was 250,000 and in 1961, it was more than 400,000, which represented 7.5 per cent of all French Canadians. The gains from immigration do not compensate for these losses. Further, this loss is not uniformly distributed and is most serious in those provinces where the French presence is weakest. In 1951, in the very British province of Newfoundland, only 24 per cent of Canadians of French origin defined themselves as French-speaking. In 1951, in the town of Cornwall (Ontario), French Canadians formed 42 per

5. Public statement made by General C. de Gaulle, Montreal, July 1967.

cent of the total population, but only 27 per cent were French-speaking.

Massive Population Growth

The fact that, out of less than 10,000 French immigrants who landed in North America in the seventeenth and eighteenth centuries, there arose a population of 6.3 million (more than 8 million, if those who emigrated are included) represents a very remarkable demographic increase. French Canadians form the second-largest group of French origin on the world scale, and Montreal prides itself on being the most important French-speaking city after Paris. The existence of this large group is a phenomenon which is largely ignored, as much by English Canada (as the evidence to the Royal Commission on Bilingualism shows) as by France itself.

In 1763, within what are now the boundaries of Canada, the French-speaking group (French Canadians, French from France, and Acadians) totalled almost 80,000 inhabitants. After 1790, the French Canadians formed a larger group than the native peoples. The number of resident French Canadians reached 1 million at the time of Confederation (1867), 2 million in 1910, almost 3 million in 1930, and 1 million more fifteen years later. The "revenge of the cradle" has therefore been astonishing, at least in quantitative terms.

For survival to be assured, it is certainly necessary to provide a generous number of births; it also requires cultural perseverance. The factors which have prevented the fusion of the great majority of individuals into the Anglo-Saxon group include physical and psychological separation, the use of a common French language, a particular way of life and a very French attitude of resistance to assimilation. Moreover, according to A. Siegfried (himself a Protestant): "Without the powerful Catholic clergy, the French Canadian would not have survived,"[6] an opinion which, to this author, seems exaggerated.

The ratio of the group of French origin to the total population has varied. Because of the early numerical predominance of Indians, it was only 5 per cent at the end of the seventeenth century. In 1760, about 30 per cent of the inhabitants residing in what is now Canadian territory were of French origin. At the start of the British period, this proportion rose to a maximum of around 40 per cent. Soon, however, British immigration was able to get the better of the birthrate and French persistence, and the proportion fell back to 26 per cent shortly after Confederation (29 per cent, if only the four provinces in Confederation are included). Since then, the proportion of Canadians of French origin has shown slight variations, caused by immigration of people from elsewhere. For two centuries—or since the British victory—the ratio of people of French origin to the total population has therefore not changed markedly; this relatively static ratio for such a prolific population helps one to appreciate the counterbalance provided by immigration of other ethnic groups. During the course of the next generation, the present fall in the French-Canadian birthrate, combined with immigration by other groups, could cause the proportion to fall below 25 per cent.

Of 5,700,000 Canadians who speak French, 3,500,000 do not know any other language. This linguistic fact creates a nation (*pays*), at least in sociological terms.

Four French Canadas

Seen in historical perspective, the distribution of French-speaking peoples can be considered

6. A. Siegfried, *Le Canada, puissance internationale*, p. 57.

in four main stages, which have similarly led to four successive French Canadas.[7]

(1) *Before 1760*. Before 1760, the first French Canada—New France, or simply, Canada—extended in a linear fashion from the Atlantic to the Rockies and to the lower Mississippi. The area was sparsely settled and was hemmed in by the British in the southeast and by the Hudson's Bay Company in the northwest. The Indian population which lived in this part of Canada was, for a century, more numerous than the French population. In 1760, French Canada was still Franco-Indian. We shall not enter the open dispute among historians on the extent to which French Canada had already formed a nation. Whether it actually was or not, the French defeat in 1760 created a second French Canada, this time an orphan.

(2) *From 1760 to about 1840*. In terms of land area, this French Canada was made smaller by the retreat sketched out in the 1763 proclamation of George III, despite the extension given by the Quebec Act of 1774. The revolutionaries in the United States undoubtedly gained the south shore of the Great Lakes. Later, the colonization of Loyalists in Ontario beyond the Ottawa led to the subdivision of this first Province of Quebec into an Upper Canada (Ontario) and a Lower Canada (Quebec); the French-speaking country was thus reduced to a sector of the St. Lawrence Lowlands. It became an enclave in which the principal ecumene was broadly coterminous with the area occupied by the postglacial Champlain Sea. On the cultural side, the return to France of about 2,000 people after the defeat and the lack of direct contact with the home country made the French Canadian more definitely *Canadien*.

Because of the reduction in area of the lands of the French-speaking, the Indian group was less numerous than before 1760; in contrast, the British soon established themselves as the second-largest ethnic group. Within Quebec, the French-speaking Canadians never constituted less than 75 per cent of the total population. This second French Canada, which lasted until about 1840, was inward-looking. Such seclusion assisted its consolidation and population growth; while becoming more numerous, the French Canadians kept their language and their religion (which had become very Roman) and developed their political institutions. In doing so, they forged a country, the defence of which against the United States assured the survival of the whole of Canada (1812).

In several ways, the British political victory in 1760 was only a partial one. Though Lord Durham, around 1840, was to regret the failure of assimilation in Quebec, the British victory was neither demographic, cultural nor religious. The "Old French", apart from several thousand Acadians,[8] were not obliged to leave their lands and, in Quebec, they have not been overwhelmed by the English-speaking element. But the victory was successful in terms of business. Under the direction of the British merchants of Montreal, the fur trade, and later the exploitation of forest resources for European markets became particularly prosperous. As H. Neatby wrote, "The St. Lawrence community became

7. J. Hamelin, "The Historic Development of French Canada", *Quebec Yearbook, 1966–1967* (Quebec, 1967).

For a study of the evolution of French Canada during this period, the following are recommended:

(1) L. Gérin, *Le type économique et social des Canadiens. Milieux agricoles de tradition française* (Montreal: Fides, 1948);
(2) J. C. Falardeau, ed., *Essays on Contemporary Quebec* (Quebec: Laval University Press, 1953);
(3) E. C. Hughes, *French Canada in Transition* (Chicago: University of Chicago Press, 1943).

8. R. A. Leblanc, "The Acadian Migrations", *Cahiers de Géographie de Québec* 24 (1967): 523–541.

an integral part of the Empire, a branch business. . . ."[9] Admittedly, this situation tended to coincide with a certain French-Canadian philosophy, according to which French-speaking Québécois are not born to be salesmen. On its part, the Church was concerned with the salvation of souls, and did not favour discussions of the economic situation. The French Canadians thus came to be a group which was hard-working, conformist, lacking a secular elite, resigned, tolerant, and very rural (farm workers, woodcutters, log drivers, small shopkeepers or coureurs de bois). In the narrow confines of the St. Lawrence, however, the opportunities for work (which was fairly undiversified) could not provide a better standard of living. F. Ouellet has demonstrated the slowness of the economic development of the French-speaking population in Quebec during this period.[10] This situation seems to justify the fatalism perceived by the historian A. R. M. Lower: "French Canadians found happiness in life, not in things."[11] However, a third French Canada had already begun to take shape.

(3) *From 1840 to about 1950*. Even before the middle of the nineteenth century, the inhabitant of Quebec tended to leave his small and overpopulated plain. Through this emigration, the French fact again ceased to be the affair of a single province; there was a progressive transplant of French Canadians to all parts of Canada, and even to the United States. By 1871, 150,000 were already living outside Quebec, half of them in Ontario. The 1961 census showed that the number living in other provinces was 1,300,000. The chief stages in this advance began with the takeover of the English Protestant townships of Quebec which, after the eighteenth century, surrounded the fertile areas of the St. Lawrence.[12] This extension penetrated to the townships bordering the former Upper Canada. Next, the French-Canadian groups, their priests in the vanguard, moved out of the St. Lawrence into the Saguenay–Lake St. John basin (after 1838) and, in the twentieth century, into the Quebec and then the Ontario Abitibi area. During this time, the former Acadian nucleus was revived and firmly established in New Brunswick, where its proportion of the total population grew from 16 per cent in 1871 to 40 per cent in 1961. In the 1960–1970 decade, the province twice elected an Acadian as premier. The dispersed part of French Canada thus came to consist of a number of large areal units and a multitude of small, scattered, and sometimes culturally overwhelmed outposts, whose distribution across the provinces seemed to attenuate with increasing distance from the Quebec core. As compared to the earlier period, three social groups were of great importance: the professions, the shopkeepers and the clergy. While these spatial and social changes were taking place, Quebec, through Confederation in 1867, became the seat of a provincial parliament, a government which had its direct origin in the 1791 constitution of Lower Canada.

This French Canada, though, was only a country in part. It can still be considered this today, since it neither has nor completely controls any of the four essential and basic attributes of a country. Politically, it shares its powers with another culturally distinct government in Ottawa. Economically, it does not control the means of production. Spatially, it does not constitute a continuous territory, since there are units separate from Quebec over which it has no jurisdiction; nor is it a continuous

9. H. Neatby, *Quebec. The Revolutionary Age. 1760–1791*, p. 4.
10. See F. Ouellet, *Histoire économique et sociale du Québec, 1760–1850.*
11. See A. R. M. Lower, *Colony to Nation. A History of Canada.*
12. See P. B. Clibbon, *Land Use Patterns in the Laurentides between the Saint-Maurice and Rouge Valleys (Quebec)*, Ph.D. thesis (Quebec, 1968).

ethnic unit, in view of the powerful English-speaking minority in Montreal. Linguistically, the language it speaks is controlled by France.

(4) *After 1950*. Since 1950, French Canada has entered a fourth and "irreversible" stage focused on comprehensive development. Among the many who seek a better country, the advocates of political separatism or independence are the most obvious. Urged on by the many young people who, in turn, are encouraged by the principle of nationalism and the example of those in the Third World who have recently been freed from the colonial yoke, Francophone Quebec reopened the question of its adherence to a Confederation which is felt to have been Anglophone, whether consciously or not. This new wave of independence (shown in June 1957 with the foundation of the Alliance Laurentienne in Montreal) came from a double awareness: first, that French Canada, in terms of social psychology if not in law, has never ceased to exist and, secondly, that it has not expanded sufficiently in the basic areas of the arts and economic affairs.

From the cultural viewpoint, the French-Canadian elite believes that the language, particularly the working language, still needs improvement, despite the massive progress achieved in the last twenty years. The problem of French-Canadian culture is becoming less and less one of identity, but remains one of expression. This does not mean, of course, that the few old-fashioned expressions the language contains are sufficient to make it a language of the seventeenth century, nor does it mean that the French of France and French Canadians—like many North Americans—do not understand each other. Further, it is undeniable that most French Canadians do not attach much importance to the need to speak and write well. Their carelessness is apparent in structure, vocabulary and pronunciation, as well as in accent. Some argue that Quebec should stop being officially bilingual, or that French should at least become the main language in industry, business and advertising. On the other hand, French-Canadian vocabulary is certainly much richer, in many respects, than that of France, since it has developed to include expressions for Canadian environmental phenomena (snow, ice, northern characteristics) and cultural phenomena (technology, sports) not found in France.

In the area of economic life, French Canadians, who constitute 30 per cent of the Canadian population, control only 10 per cent of the national economy. Even in Quebec, where they constitute 80 per cent of the population, they control only 25 per cent of business. This is not something which has developed recently, however. Such statistics imply that economically, the French Canadians are still in a colonial state. Some progress has been made in insurance, banking and some forms of industry—the powerful Desjardins financial group, for example. It remains true, however, that the French Canadian is often only a simple participant found at the lower levels of the economic hierarchy.

A growing self-awareness means that the people of Quebec can choose among a variety of political formulae, including federalism (revised to recognize special status for Quebec), a true confederation of associated states, or independence. At the end of 1967, René Lévesque announced his objective: sovereign status for Quebec and association with English Canada. Quebec independence is not a new idea; by 1880, some French Canadians had begun a psychological withdrawal from the 1867 Confederation. Some regard the present political separatist movement with sometimes violent enthusiasm; others see only a new manifestation of a periodic "emotion", directed inwards. Be that as it may, the heartland of French Canada is preoccupied with its cultural consolidation. There is talk of "The Quebec State".

An outward-looking Quebec is assuming a supraprovincial role by helping the other French communities of Canada. The French-Canadian Association for the Advancement of Science (A.C.F.A.S.), previously a Quebec institution, opened a branch in Moncton in 1967. Externally, Quebec wishes to be recognized as a distinct member of the French-speaking world; in 1968, the province participated in the Gabon conference. French Canada can perhaps be extended formally outside Quebec, if Ontario and New Brunswick accept the suggestions of the Commission on Bilingualism and Biculturalism.

The future of French Canada involves major problems: the improvement of written and spoken French; the transformation of the Quebec economy for the benefit of those who speak French; the fate of the French-Canadian minorities in the provinces of the extreme east and in the west of Canada; the economic contrast between the very large, powerful and developed Montreal and the rest of Quebec, which is largely underdeveloped; the development of the North; and political relations with the rest of Canada, which run the risk of break-up, and with the United States, a militarily-nervous neighbour and provider of capital. Independent or not, Quebec must find solutions to these problems, and not for its own sake alone; for it is part of an integrated North American economic, social and cultural system.

Canadians of British Stock

From one historical viewpoint, English Canada (by descent) forms the other foundation on which the political edifice of Canada has been built. It has certainly been the major support of the country for more than a century. It might even be considered the only one, if we were to take literally the official title of the 1867 Constitution, the *British* North America Act.

English Canada consists of two groups: those who are of British descent, and the Canadians of all kinds for whom English has become the language regarded as the mother tongue. The difference between these two ways of being an English Canadian is important; whereas French-speaking Canadians were rather fewer than the Canadians of French descent, here the opposite is true, and to a marked degree. In 1971, there were about 9,000,000 English Canadians by descent but more than 12,973,810 by language, emphasizing the assimilating effect of the English language. English Canada is, therefore, more English-speaking than strictly British. We are concerned here with the Canadians of British descent, who comprise several subgroups:

(1) the oldest groups, the Canadian identity of which is well established. These include: (i) the descendants of eighteenth-century immigrants to the Maritimes and (after 1760) to Quebec; (ii) Loyalists from the United States; and (iii) British who, in the second quarter of the nineteenth century, gave real meaning to the existence of Ontario and even the Maritimes

(2) an intermediate type, composed of the grandchildren of the early twentieth-century immigrants who settled on the Prairies

(3) some young groups such as: (i) the post-war immigrants who have mainly swelled the numbers in urban areas; and (ii) U.K. citizens or those from other Commonwealth countries who stay for several years before moving on, stage by stage, to other English-speaking countries

Superimposed on all these groups are the differences between the inhabitants of various parts of the British Isles; differences, for example, in the life styles and even the religions of the English, Scots and Irish. In 1961, the English in the strict sense constituted a little more than half the British in Canada; this group is powerful politically and strong in the preservation of its traditions. It is worth noting that these 4.5 million individuals do not equal the number of French Canadians. Second largest are the Scots (23 per cent of the population of British descent), who have played an important financial role in Canada. They were the founders of the North-West Company, the Bank of Montreal and other economic institutions such as insurance companies. Lastly, the two parts of Ireland comprised 22 per cent of the British Isles' representation in Canada. Nevertheless, though all the preceding distinctions were real, integrating elements such as language and business provided a social unity to these groups, resulting in an impressive total of 8 million individuals. Within the total, those born in Britain should be distinguished from English-speaking Canadians born in Canada. In certain respects, this distinction may be difficult to draw but in other ways—speech, social attitudes, favourite sports—the difference is apparent.

General Evolution

Compared to the indigenous peoples and to the French, the British group was slow to grow. Although in 1627 the single English colony of Ferryland (Newfoundland) "was as important as all of New France",[13] it was only at the end of the first quarter of the eighteenth century that the British reached 10,000 inhabitants, almost half a century after New France had grown to that size. Though French and English were virtually contemporaries in the "discovery" of the Atlantic coasts of Canada, the French were in the lead in the areas of colonization, settlement and the number of inhabitants.

At the time of the French defeat, English Canadians still represented only 12 per cent of the total population of Canada; their military success can only be explained by the direct assistance of the British navy and by the immense human and psychological reserve represented by the thirteen colonies, still politically British. It was not until the beginning of the nineteenth century that the English-speaking peoples rapidly grew in numbers, exceeding first the native peoples and then the French Canadians (see table 12.1). Once relieved of the burden of European affairs, Great Britain sent out large numbers of emigrants; from 1815 to 1865, it seems probable that 750,000 British arrived. Canada, accustomed to the grudging release of immigrants, changed its appearance totally; it grew in size, became of economic significance, developed links abroad and, most of all, it became British: "The principal bond between all the British communities is in that first word of the collective title *British* North America: they were *British* in contradistinction with *American*: monarchists, not republicans."[14] By the middle of the century, the British element already represented 65 per cent of the total population. At the time of the conference which led to Confederation, there was hence no real demographic justification for political equality between the two groups, French Canadians forming only 26 per cent.

After 1860, emigration from Ontario and then the Maritimes caused the proportion of Canadians of English descent to diminish. During the first three decades of the twentieth century, British immigration again took place, but so strong was the immigration of some other European groups that the proportion of inhabitants of British descent in Canada declined even more. Since 1941, they have formed less than 50 per cent of the total population; today they represent about 42 per cent.

13. M. Trudel, *Histoire de la Nouvelle-France*, Vol. 2, p. 413.

14. See A. R. M. Lower, "Canada at the Turn of the Century, 1900", *Canadian Geographical Journal* LXXI (1965): 3.

TABLE 12.1 Proportions (%) of the Main Ethnic Groups in the Canadian Population, 1600–1961

Date	*Indians and Eskimos*	*French Descent*	*British Descent*	*New Canadians*
1600	100			
1680	93	5	1	
1760	57	29	12	
1790	36	36	27	
1810	22	39	39	
1830	11	37	51	
1850	4.7	27	65	2
1871	3	26	63	7
1901	1.8	29	59	9.5
1921	1.3	27	56	14
1961	1.2	30.4	43.8	24.6

SOURCE: Calculated from census data.

NOTE: For the years before 1850, the proportions are tentative estimates.

Other Characteristics

In terms of provinces, English Canada is more extensive than French Canada. It is, however, in the minority in two of the five major regions: in French-speaking Quebec; and on the Prairies, where ethnic groups other than English, French and Amerindian are slightly more numerous (table 12.2).

Some provinces are very British, such as Newfoundland, a Crown colony until very recently. In Quebec, by contrast, the 500,000 British represent only 10 per cent of the total population. In British Columbia, as in Ontario, the British proportion does not exceed 60 per cent because of the increase of immigrants of all nationalities, including Germans, Italians and even French Canadians. Altogether, 70 per cent of the inhabitants of English Canada live east of Lake Superior, and the proportion would be still higher in the case of French Canada. The majority of members of these two principal peoples are always to be found side by side along the St. Lawrence axis, from the Great Lakes to the Gulf.

Since Canada is officially defined as biethnic, bilingual and even bicultural, it is appropriate to look at some of the quantitative relationships between French Canadians and English Canadians. These relations have varied inversely over time. For every 10 persons of English origin, there were 19 French Canadians in 1760, no more than 4 in 1850, but 7 in 1961. In terms of language, French Canada has scarcely made any relative progress during the course of the last century, because the great majority of other immigrants have chosen English as their mother tongue. Thus, for every 10 who spoke English in 1850, there were only 4 who spoke French, and less than 4.5 in 1971. The linguistic gains achieved by English Canadians among other cultural minorities compensated for the relative decrease in those of British descent.

This has not been the only form of help. In marked contrast to the French element, those

TABLE 12.2 Distribution of Ethnic Groups by Province, 1961

Area	*British Descent*	*French Descent*	*New Canadians*	*Indians and Eskimos*	*Total*
Newfoundland	428,899	17,171	10,372	1,411	457,853
Prince Edward Island	83,501	17,418	3,474	236	104,629
Nova Scotia	525,448	87,883	120,405	3,271	737,007
New Brunswick	329,940	232,127	32,948	2,921	597,936
Atlantic Provinces	1,367,788	354,599	167,199	7,839	1,897,425
Quebec	567,057	4,241,354	429,457	21,343	5,259,211
Atlantic Provinces and Quebec	1,934,845	4,595,953	596,656	29,182	7,156,636
Ontario	3,711,536	647,941	1,828,541	48,074	6,236,092
Eastern Canada	5,646,381	5,243,894	2,425,197	77,256	13,392,728
Manitoba	396,445	83,936	411,878	29,427	921,686
Saskatchewan	373,482	59,824	461,245	30,630	925,181
Alberta	601,755	83,319	618,316	28,554	1,331,944
Prairies	1,371,682	227,079	1,491,439	88,611	3,178,811
British Columbia	966,881	66,970	556,417	38,814	1,629,082
Western Canada	2,338,563	294,049	2,047,856	127,425	4,807,893
Yukon and N.W.T.	11,725	2,403	8,058	15,440	37,626
Western Canada and Territories	2,350,288	296,452	2,055,914	142,865	4,845,519
Canada	7,996,669	5,540,346	4,481,111	220,121	18,238,247

SOURCE: 1961 Census.

of English descent form a group which is backed up by a homeland concerned with it, thus providing immigrants, ideas, administrators, trade relations and, for a long time, the majority of the capital and equipment. How different from the French group, which was so early separated from its mother country by the hazards of war. This difference, which has persisted for two centuries, explains much of the divergence in population and power between the two founding elements of the country. While the English Canadian benefited from the powerful magnetic field of the British Empire, the French Canadian could only rely on his own inspiration. Further, whereas the former is still partially British, the latter is no longer "French of France"; conversely, though both are North Americans, the English Canadian is more recognizably American.

Two centuries of living together have scarcely succeeded in achieving a significant bond between those of English and those of French origin. As in Europe, a psychological barrier hinders communication between the two groups; the distance remains great and the difficulties of the journey across are real, as the historian Lower has recently written. In the nineteenth century, the British influence was so enormous that French Canada was forgotten. It is significant that in Vancouver a symposium devoted to Quebec's cultural separation gravitated naturally towards a simple economic regionalization of Canada. Only fifty Canadians of British origin out of a thousand can express themselves in French. English Canada developed after, beyond and above French Canada. There are thus two Canadas, called "two solitudes" by some. The ethnic difference, emphasized by religious and spatial separation, thus continues to distinguish French Canadians from English Canadians. While not considering each other total strangers, they scarcely meet. With some exceptions, these two worlds remain nearly as separate as the French of France and the British in Europe. However, curiously enough, this sociological separation does not usually lead to antagonism.

Other Groups: The New Canadians

The title of a perceptive book by André Siegfried, *Canada, les deux races*, is out of date.[15] When it was written, at the beginning of the twentieth century, the two major cultural groups in Canada made up 90 per cent of the Canadian population. At the 1971 census, however, more than 2.5 million people had a mother tongue other than English or French. In 1964, 104 papers and periodicals appeared in languages other than the official ones. In Vancouver, Christian prayers are offered in seven languages. Through this evolution, Canada, which has become scarcely more bilingual, has also ceased to be biethnic. For convenience, we shall use the term *others* to describe the group composed of those whose origin is not British, nor French, nor native. They are also often called *New Canadians* because of their recent immigration.

Development of Ethnic Diversity

Apart from some individuals and some isolated settlements, such as the Germans in Nova Scotia in the middle of the eighteenth century, it was only during the first quarter of the nineteenth century that the "other" groups passed the 10,000 threshold, about a century after the British and almost a century and a half after the French. At Confederation, around 7 per cent of the inhabitants came from such varied descent. The abundance of ethnic groups is thus a characteristic only of the twentieth century: the proportion grew from 9.5 per cent in 1901 to 14 per cent in 1921 and to more than 25 per cent in 1971. Ten years ago they represented 4,500,000

15. See A. Siegfried, *Le Canada, les deux races* (Paris: Colin, 1906).

inhabitants, only 1,000,000 or so fewer than the French Canadians themselves. This "foreign" sector already represents 60 per cent of the size of the British element, and its natural rate of growth and its rate of immigration are higher than those of the British. From 1954 to 1963, only 27 per cent of the immigrants to Canada were British, the majority of foreigners coming from non-French-speaking areas of continental Europe. This polyethnic group has not yet displaced the earlier groups in relative importance, as the French did the native peoples and later as the British did the French Canadians.

Characteristics

The contributions of different nations to this ethnic conglomeration are far from equal in size. At the 1870 census, of the 283,000 "others", the Germans overwhelmingly dominated with 203,000; far behind them came the Dutch (30,000) and a dozen other nationalities. During the period 1900–1915, the largest contingents of immigrants (over 100,000) came from Russia, from Poland and from Italy. Groups of about 50,000 left Germany, Scandinavia, the Balkans, China and Japan. There were about 47,000 Jews and the statistics record immigrants from seven other countries. During the ten years preceding the 1929 Depression, the greatest number of immigrants to Canada came from the U.S.S.R., Germany, Scandinavia and the Balkans. By contrast, Italy, which was preparing for fascism, and Poland, enjoying a recovery, allowed fewer of their citizens to leave. Since 1946, Germany, Italy and the U.S.A. have become the most important sources; some countries with small populations, such as Greece and Portugal, have been close to the leaders among the twenty-eight other countries. Only a proportion of the immigrants remain, and hence the ethnic composition of the resident population does not precisely reflect the rates of immigration. At the 1961 census for Canada there were: Germans, 1,049,599; Ukrainians, 473,337; Italians, 450,351; Dutch, 429,679; Poles, 323,517. Seven other groups, including Jews, exceeded 100,000. The German Canadians have formed the most significant group for more than a century.

Very varied ethnically, these New Canadians, like the British and French Canadians, have areas where they are well established. The Prairies are the most important part of their ecumene; here the New Canadians represent 46 per cent of the total population, rather more even than the British, and they have contributed greatly to development. The second, scattered home is urban Canada: Vancouver, Montreal, Halifax, and particularly southern Ontario. This has resulted in an approximately equal balance of "others" between eastern and western Canada.

In more detail, the areas where the different groups concentrated were at first very distinct from one another. In cities, most of the islands of national culture have been assimilated and the tendency in recent times to live in areas according to income rather than according to family tradition and kinship has diminished the importance of the urban ghettos.

The admission of immigrants raises the question of their assimilation, a very difficult problem to study or even to define. There is, in fact, no universal Canadian archetype against which the distinctiveness of New Canadians can be judged. Canada is a young country developing from more than one ethnic group; further, it has not known lengthy periods of isolation from external events, during which the peoples of Europe acquired their distinctiveness. There have not been enough outstanding events which might have led each group to develop some strong feelings of nationality. A Canadian identity is being created but it is not defined in the same way by English Canadians as by French Canadians. This being the case, we have no single standard for comparison.

Certain things tend to slow down the assimilation of immigrants, these including: a tendency to live in national groups centred on a church and encouraged by newspapers, as exemplified by the Italians; religion (the Jews); the resistant psychology of the French of France; the lack of social pressure to assimilate; and the physical distance which limits contact. There are, however, factors which encourage assimilation into the English-speaking group. They include: the ethnic origin of a great number of immigrants (Anglo-Saxon); the ubiquity of North American material culture; the superstructure of federal power; the highly developed physical and cultural communications (including television); the mobility of the workers; and a tendency to marry outside the cultural group.

Legally, the first stage of assimilation is citizenship; 94 per cent of the population is Canadian legally and in fact. Strong postwar immigration which has continued to the present day is the reason why the proportion of aliens has risen to 6 per cent: a minimum of five years is required to acquire the legal status of a Canadian citizen.

Assimilation is mainly an anglicization developed in two stages: English first becomes the normal day-to-day language, and then for the next generation it is the mother tongue. The fact that English was the basic language at the time of the settlement of the West, and even more the predominance of English in North America, has benefited this language at the expense of all others, even the immigrant's actual mother tongue. In 1961, 90 per cent of the "other" Canadian residents who had been in Canada for more than three years could speak English, while only 5 per cent could speak French. Far from winning in terms of language, French Canada has scarcely maintained its relative position. The "others" thus make only a modest contribution to the official policy of a bilingual Canada. The rate of anglicization varies among the national groups; the Germans and the Dutch adapt quickly, the Slavs (Ukrainians), Latins and Asiatics more slowly.

Although there is no official segregation of any kind, some New Canadians would argue that it is more difficult to succeed in Canada if one lacks a British or even a French ancestry. This opinion reflects a dilemma which has never been clearly faced; on the one hand, Canada declares that its doors are open without cultural distinction, while on the other hand, at least until very recently, it tried to define itself as only bicultural. It is easy to see why the adjective *national* does not have the same overtones in Canada that it can have in a country characterized by only one culture. To live in Canada, to acquire Canadian citizenship, or to be sociologically one type of Canadian or another—these remain for many people quite distinct things.

In Canada, apart from the Prairies and the major urban areas, assimilation is more limited than it was in the United States, where the "melting pot" was an accepted policy goal. Canada is more tolerant than its big neighbour.

This group of "others" is of considerable importance to Canada. Its high birthrate helped to populate the Prairies, and the same factor is still significant in the big cities. It has provided great numerical support to those who speak English. The influence of New Canadians is undeniable and they have already provided a prime minister for the country. They challenge the duality of Canada: two ethnic groups, two languages, two religions, and even two big provinces (Ontario and Quebec). Not wishing to be considered second-class citizens, these "others" lodged official protests against the Royal Commission on Bilingualism.* If a better cohesion were achieved among these thirty nationalities, which is unlikely, they could force Canada to redefine itself in quite different terms than now exist.

Conclusion

In Canada, the English aspect dominates in terms of national descent and even more in

* The first book of the report was published in 1967.

TALE 12.3 Major Groups by Descent (1961) and by Language (1961 and 1971)

Group	*By Descent**		*By Mother Tongue*				*By Ability to Speak Language**
	1961		*1961*		*1971*		*1961*
	Number	*Percent*	*Number*	*Percent*	*Number*	*Percent*	*Number*
English	7,996,669	43.8	10,660,534	58.4	12,793,810	60.2	14,515,934
French	5,540,346	30.4	5,123,151	28.1	5,793,650	26.8	5,721,038
Indian and Eskimo	220,121	1.2	166,531	0.9	179,825	0.8	
Others	4,481,111	24.6	2,288,031	12.6	2,621,020	12.2	

*Data for 1971 not yet available.

SOURCE: 1961 and 1971 Census.

terms of language. There is a total of about 16 million who are English-speaking (i.e., able to speak English) but those of British origin do not amount to 60 per cent of these: hence it is by language rather than by race that Canada is an English country. Conversely, on the French side the language is almost entirely a matter of descent. Apart from these two groups, if one takes the culture mixture which forms under 30 per cent of the total population, this heterogeneous group uses English overwhelmingly. In short, the Canadian population is polycultural in origin, not very bilingual and generally likely to speak English. The native element, even though it was in the majority during the first two centuries of colonization, scarcely counts any more.

The phases of demographic evolution indicate that Canada became more "white" than native in the eighteenth century, more British than French-speaking in the nineteenth century, more English-speaking than of English stock in the twentieth century, and now is more of "other" origins than of French origin (see table 12.3).

This multicultural society has its limits. Canada, a northern country without plantation agriculture, was spared the forced importation of black workers, the plague of European colonization in tropical America. It has not, however, been completely free of slavery; Marcel Trudel carefully documented a total of 3,600 slaves, of whom the majority were Amerindians.[16]

Negro slaves, dying very young, had little influence on population growth. This fact, added to a rate of African immigration which has always been low, explains why at the 1961 census blacks numbered only 32,000. In this respect, Canada is very different from the U.S. Asiatics, whose entry has scarcely been encouraged, represented only 0.7 per cent of the total population. It would seem likely that the number of Africans and Asiatics will not grow very much, since those with the skills required for admission to Canada are precisely those who are needed and lacking in the Third World. Canada, therefore, remains a white man's country, a projection of Europe.

Lastly, this analytical study of ethnic groups shows the complexity of the Canadian society. If such a structure complicates the development of a national feeling, it does not prevent it. More and more citizens, ethnically different, are proud to live in Canada, to benefit from Canadian institutions and resources, to be Canadian.

16. See M. Trudel, *L'esclavage au Canada français* (Montreal: Laval University Press, 1960).

Chapter 13

The Labour Force

The geography of employment provides an ideal transition from a study of population to that of the Canadian economy, since members of the population are simultaneously producers and consumers of goods and services. The size of the labour force is calculated for statistical purposes as a proportion of the active population.

In step with the growth of total population and the development of the country, the total number of workers has increased greatly since the beginning of the century, rising from under 2 million to more than 8 million. The proportion of people in employment varies according to immigration, economic conditions, birthrates, the age structure and the female participation rate. In 1970, the labour force represented

55.8 per cent of the population aged fourteen and over. The United States percentage would be a little higher, due to a greater number of women in employment.

Increasing Dominance of the Tertiary Sector

A useful way to follow national development is to examine the range of activities of its inhabitants at different periods (table 13.1). Despite changes in statistical definitions and even in the character of the employment itself, this type of study reveals some fundamental characteristics.

The relative importance of the first and the last of the four categories has changed markedly. Those employed in agriculture, hunting, fishing, mining and lumbering represented half the labour force at the beginning of the century, but now form the smallest group. In 1971, only 6.5 per cent of the labour force was engaged in agriculture. This decrease reflects the static character of the agricultural area in contrast to other types of work which are growing rapidly. In 1961, the agricultural labour force of 648,000 was about the same as the number employed in the professions and it was greatly exceeded by the number of industrial workers (1,036,000). All the major regions of Canada have recorded declines in the primary employment sector. It has been estimated that depopulation from prairie farms during the 1951–1961 decade was about 200,000. On the Atlantic coast there has been a similar decline in fishermen. From 1901 to 1951, the percentage of primary jobs as a proportion of total employment fell from 18 per cent to 12 per cent in British Columbia, from 60 per cent to 36 per cent in the Prairies, from 42 per cent to 12 per cent in Ontario, from 40 per cent to 16 per cent in Quebec, and from 49 per cent to 24 per cent in the Atlantic Provinces.[1] The present proportion of employment in the primary sector could decline even more, since the proportion of nonresident farm owners in Quebec is as high as 51 per cent.[2] Despite the importance of Canadian wheat on the world market, Canada today is not really an agricultural country.

If that is true, can Canada now be considered an industrial country in terms of employment? Not by dominance. Although the category of manual workers (including workers in manufacturing industry and construction and unskilled labourers) forms the second-largest group, it comprises barely one-quarter of the labour force. Rather surprisingly, it has represented an almost constant proportion of the total, because the growth of blue-collar workers has been matched by that of white-collar workers. Around 1950, however, the subcategory of manufacturing displaced agriculture to become the leading single source of employment.

In reality, in terms of most people's jobs, Canada is neither a country engaged in the production of raw materials nor a factory in which they are processed. If we take tertiary and quaternary employment together, the last seventy years have seen these sectors grow from 23 per cent of the total to 60 per cent. In these terms, Canada is really no longer a new country. Whereas at the beginning of the century three workers out of four were occupied in producing goods, around 1951 the services group began to dominate. The category of "brain workers" has grown particularly rapidly and has moved from last place to first place. This spectacular rise, related to governmental bureaucracies, urbanization, the growth in female employment, and the development of intellectual activity and industrial management, is an indicator of economic and social maturity.[3]

1. See [Canada], *Resources for Tomorrow*, Vol. 3.

2. See [Quebec], *Atlas du Québec, L'agriculture*, compiled by the Ministry of Industry and Commerce (Quebec, 1966).

3. See N. M. Meltz, *Manpower in Canada 1931–1961*, report by the Department of Manpower and Immigration (Ottawa, 1969).

TABLE 13.1 Canadian Employment, 1901–1961

Sector	1901		1961	
	%	*Order*	%	*Order*
Primary	50	1	13.1	4
Secondary and construction	27	2	27.1	2
Tertiary and personal services	16	3	26.4	3
Government, the professions and office employment	7	4	30.8	1
Unspecified			2.6	
Total	100		100.0	

SOURCE: Compiled from census data.

During the twentieth century, therefore, Canada has undergone very significant changes. The classic Canadian types of employment—the coureurs de bois, the trappers, the voyageurs, the lumberjacks, the raftmen, the sourdoughs, the habitants, the farmers and the cowboys, and even the housewives—have disappeared and have been replaced by those in manufacturing industry, by office workers, technicians, and by tertiary employment generally. But these rapid and very recent shifts in employment do not seem to have been accompanied by similar social adjustments; in some respects, urban society has retained a rural mentality.

Regional Variations in Employment

The overall Canadian picture obscures some distinctive features in the major regions. In the field of primary employment, there is a sharp contrast in western Canada between the granary of the Prairies and the limited amount of agriculture in British Columbia. In secondary employment, the established industrialized provinces of Quebec and Ontario have been the antithesis of the Prairies. In terms of labour productivity, the Atlantic Provinces take last place. The two most widely separated areas, British Columbia and the Maritimes, have the highest percentages of employment in the tertiary sector. Ontario, a province which contains not only its own capital but the federal capital as well, leads in the quaternary sector. In terms of occupational balance, the Prairies seem to be the most fortunate area.

The Problem of Unemployment

Involuntary stoppage of work represents one of the permanent human problems in Canada. As a proportion of the potential labour force, the unemployment figures range from 19 to 1 (table 13.2).

In economic terms, a policy of "full employment" would leave about 3 per cent of the labour force unemployed. In relation to this threshold figure, the economic difficulties of the 1929 Depression, when the rate of unemployment in the worst-hit regions went as high as 30 per cent, can be appreciated. Massive unemployment is a catastrophe not merely for the individual worker but also for the overall economy. In 1961, it was estimated that the loss of business represented a relative loss of $2 billion in national revenue.

Economists distinguish between three main types of unemployment. *Structural* unemployment represents a prolonged stoppage of work

TABLE 13.2 Percentage of Unemployment

Representative Years	*National Average*
1933	19
1939	11
1941	4
1945	1.6
1951	2.4
1961	7.2
1966	3.6
1971	6.4

related to major causes connected with, for example, the trade cycle, and with the fundamental facts of national life. This type of unemployment affects mainly eastern Canada. From 1958 to 1962, for example, the Maritimes had unemployment rates varying between 10.8 per cent and 12.5 per cent, while the Canadian average varied from 5.9 per cent to 7.2 per cent, and a proportion of this Maritime unemployment arose from structural causes.

The second type can be termed *frictional.* This consists of temporary unemployment caused by a short-term imbalance between the supply and demand for labour. The arrival of new workers on the labour market or the time required for workers to move from one province to another are examples of this minor form of unemployment.

The last category, determined by the temperature changes in Canada, is *seasonal* unemployment. Not all work stoppages due to climate lead to unemployment. For example, the advent of winter does not automatically make the harvest-worker unemployed: he may turn to other things or take his holidays. Some other workers are in a similar situation. We must therefore distinguish carefully between the annual cycle of work and seasonal unemployment patterns. In the strict sense, seasonal unemployment from 1946 to 1963 has been responsible for about 1 per cent of total unemployment, which has reached a maximum of 7.2 per cent. Of those unemployed, 40 per cent are out of work for a period of three to four months. Despite recent winter works programmes, seasonal unemployment is still apparent in some activities of the construction industry, but is much less noticeable in manufacturing.

The rate of unemployment varies from region to region. During the last twenty years, western Canada and Ontario have experienced lower rates than those in eastern Canada. For structural reasons (types of industry, pattern of economic growth, age and educational level of the labour force, mobility), the Prairies (mainly because of Alberta) have an average unemployment rate one-third that of the Atlantic Provinces.

Employment might still undergo major changes. As a result of a long period of high birthrate from the end of war until 1963, working Canada is being invaded by a mass of youth. According to the Economic Council of Canada the labour force in 1970 reached more than 8,000,000, which represents 1,700,000 more than in 1963. The Council estimates that the percentage increase in the Canadian labour force currently exceeds that of the United States, the Netherlands, Japan, France, Italy or Germany. This demographic pressure represents a challenge. It will be accompanied by a reduction in the number of weekly hours of work (more than sixty-five around 1870, forty-

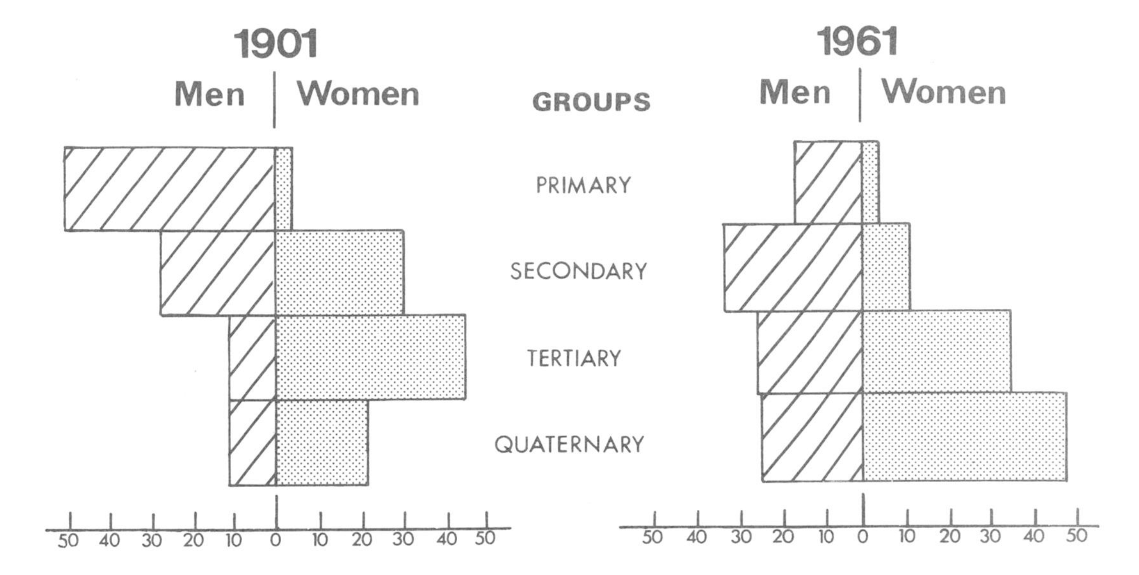

Fig. 13.1 Employment profile: Canadian manpower percentage. (From the 1901 and 1961 Census.)

five in 1950, perhaps thirty-five around 1975). Further, the increase in the total population will require more goods and services, houses, roads, schools, and recreation areas, all of which will tend to increase the size of the national market.

A second group of problems affects specific sectors of the labour force. First, the number of women will become increasingly significant in the labour market (fig. 13.1). Labour legislation must develop to accommodate this and move towards parity in conditions and salaries. It seems reasonable to forecast a growth in part-time employment.

As for the 400,000 male workers who have little technical skill, they raise very difficult problems. For the year 1960, the Economic Council of Canada calculated the relationship between workers' annual earnings (mean) and their educational level as follows: elementary schooling—$3,500; three years of high school—$4,500; five years of high school—$5,000; some years of college—$6,000; university degrees—$9,000.[4]

In a country where the categories of employment are now linked more to school education than to muscle power, the opportunities for general technical education must be greatly improved. The alternative is the deliberate continuation of unemployment and low earnings.

Changes in the birthrate since World War II may create an aged manpower situation during the last quarter of this century and a relatively small number of workers in the early years of the twenty-first century. Such basic demographic trends must be considered in any future planning.

---

4. See S. Ostry, *Unemployment in Canada*, 1961 Census Monograph, prepared by the Economic Council of Canada (Ottawa, 1968).
   See also [Canada], *Economic Council of Canada 1965*, Second Annual Review.

Part Four

Economic Organization and Activity

The strength of Canada is based mainly on its economic activity. Whether assessed in terms of capital assets or volume of business, average standard of living, per capita output or level of international trade, this country, however few the number of its producers and consumers, is among the world's leading nations. This remarkable achievement is a clear demonstration that, despite frequent setbacks, the main determining influences have been the rich resource base and the stimulus of the U.S.A. For the same reasons, the major problems facing Canadians are, to a large extent, economic ones: national ownership of capital and business, the stability and variety of external trade markets, relationships between business and government, division of economic powers between the federal and provincial governments, urban renewal, pollution, development of the North and the revival of rural areas. Such economic issues have a significant effect on Canadian political problems.

The rate of growth of the national economy has been very rapid. From 1870 to 1972, the Gross National Product multiplied more than thirty times, while the total population increased only six times. Despite a decline in the value of money, this impressive growth in the quantity of goods and services produced accounts for the present high standard of living of the Canadian population. For the first time, in 1972, the GNP reached $100 billion.

Throughout the last hundred years, though buffeted by a variety of booms and declines, Canada's economy has shifted. Previously based on a small number of primary products (staples) for export across the North Atlantic, it is now characterized by large and diversified production and consumption, pronounced industrialization and by very close ties, more through inevitability than through choice, with the economic power of the United States.[1]

Since 1945, there has been, successively, a decade of progress, five difficult years, and then a period of sustained prosperity which has continued through the problems and inflation existing since 1967. During this period, the Canadian economy grew, diversified and matured.

1. R. I. Wolfe, "Spatial Interaction in Economic History of Canada", in J. Warkentin, *Canada. A Geographical Interpretation.* See also M. Q. Innis, ed., *Essays in Canadian Economic History.*

Chapter 14

A Market Economy

This is an essential topic in the study of a country characterized by a capitalist economy and a high standard of living. Three aspects only will be discussed: investment, consumer expenditure and the balance of payments. Questions of costs, incomes, prices, financial institutions, external debt, the amount of money in circulation, the financial market and economic policies will only be touched on indirectly.

TABLE 14.1 Long-Term Investment (%) in Canada from External Sources

Year	*Great Britain*	*U.S.A.*	*Other Countries*	*Total Investments (in millions of dollars)*
1900	85	14	1	1,232
1939	36	60	4	6,913
1961	14	76	10	23,570
1967	10	81	9	34,702

SOURCE: N. K. Inman and F. R. Anton, *Economics in a Canadian Setting*, and *Canada Year Book, 1971*.

Investment: Large in Amount and External in Origin

Economically as well as politically, investment is one of the fundamental factors shaping modern Canada. On the one hand, because of the rich natural resource base, the method of exploitation and the history of the North Atlantic economy, capital has always been important: in 1955, gross capital investment per capita amounted to $418 in Canada, compared to $387 in the United States and only $151 in western Europe.[1] On the other hand, a continuing concern for Canadian independence has transformed the question of American investment into a political issue which concerns governments as much as individuals.

By following changes in the pattern of investment over time, one can observe the main characteristic: Canada's dependence on external sources. Before the middle of the nineteenth century, investments in infrastructure, settlement schemes, fisheries, furs and timber all depended on external financing. During the first great North American enterprise, that of railway building, internal sources of finance again proved insufficient. Around 1850, British capital was brought in to enable the construction of the Grand Trunk Railway, as it was later for the Intercolonial and especially for the Canadian Pacific. In turn, the railways provoked investments in related sectors of agriculture, commerce and domestic life. Subsequently, in the last quarter of the nineteenth century, a new tariff policy encouraged Canadian industry, and as a result, increased the proportion of domestic capital investment.

This capital market has been affected by major changes in the twentieth century, as capital requirements have again exceeded Canadian investment capacity. The opening of the West for settlement, new railways, metallic minerals, the national timber policy, industrial needs in wartime and the functioning of a trading economy all require large amounts of capital which inevitably have had to come from foreign sources. Every time Canada has undertaken a major enterprise, external finance has been required. Since the interwar period, the United States has become the chief source of this non-Canadian capital (table 14.1).

For a brief period, the contraction of the world economy during the depression (1929) and increasing amounts of wholly Canadian capital reduced Canada's dependence on the state of the international capital market. But World War II, consumer needs after the war, the Korean War (1950), discoveries such as petroleum (1947) and iron (1954), enterprises such as the St. Lawrence Seaway, rapid urbanization and the activity of governments in the social field all combined to produce a strong current of investment up to 1957. This was repeated from 1963 to 1966, especially in public finance (1965). If the areas of demand for new

1. See G. W. Wilson et al., *Canada: An Appraisal of its Needs and Resources*, p. 308.

TABLE 14.2 Key Canadian Industries (%) Controlled by Foreigners

Industry	*Percentage*
Automobile industry	95
Petroleum refining	90
Chemical industry	75
Mining enterprises	60
Manufacturing companies	60

SOURCE: Walter Gordon, Conference, Waterloo, June 1971.

investment during the last forty years are analyzed, two significant features are apparent. One is the importance of investments in the public sector (e.g., social legislation); the other is the growth of investments in manufacturing and in oil and gas.

It is difficult to estimate the overall proportion of foreign investment; a likely figure would be $44 billion in 1971 (see table 14.2). In certain sectors of the economy, however, the proportion is much higher. Manufacturing is an example; apart from the automobile, rubber industries and petroleum industry, the proportion is also high in iron ore, aluminum, nickel, paper and chemicals (see table 14.3). The trend is growing and new American investments increase at roughly $2 billion a year. No other industrial country has a similar degree of external control over its manufacturing output. It is the more serious because it is concentrated in sectors where the Canadian economy is powerful and largely oriented towards export.

Despite the situation in industry, the major part of the working capital of business as a whole is Canadian, including important sectors such as agriculture, retail trade, home construction, public utilities, food and clothing. Recent major increases in government investment—even though they themselves are partly financed by external loans—have helped to increase the Canadian share of total investment.

The problem of foreign investments is perennially a matter for debate, with a notable contribution being made by the Gordon Report

TABLE 14.3 Foreign Capital Invested in Canada – 1930, 1963 and 1967
(in millions of dollars)

	1930	*1963*	*1967*
Manufacturing	1,459	7,074	10,017
Petroleum and natural gas	150	4,703	6,009
Government securities	1,706	4,207	5,813
Financial	543	2,847	3,415
Other mining and smelting	311	2,347	3,150
Public utilities	2,878	1,830	1,830
Total (all categories)	7,614	26,203	34,702

SOURCE: *Canada Year Book* for 1966 and 1971.

in 1957.[2] Despite Canadian investments abroad (mainly private investment in the United States and government funds destined for the Commonwealth), which in 1970 reached more than $21 billion, there remained a deficit in the external market of $23 billion (the deficit was $26 billion in 1967). The dilemma that frequently arises involves a choice between expansion associated with foreign capital, which increases production, employment and revenues, or near stagnation due to lack of capital. In the former case, theoretically, the economy is lively but dependent; in the latter, it is sluggish but autonomous. Since foreign capital is sometimes a necessary evil—for example, in the development of some mineral resources—various measures might be taken that would enable Canadians to participate in foreign-owned businesses located in Canada and to align American-owned enterprises more in accordance with national economic interest. A greater variety of foreign-capital sources is equally desirable; instead of dependence on the U.S.A. and Britain, greater amounts might be invested by the Netherlands, Belgium and Switzerland and especially by France and Japan.

An improved investment policy might have three objectives:

(1) to enlarge overall availability of capital, facilitating the growth required to provide jobs for the young people who are now arriving on the labour market

(2) to reduce the annual fluctuations in the level of total investment, particularly in regard to investments in specific sectors of the economy

(3) to fight regional disparities

2. See W. L. Gordon, *Notes for Remarks*, report to the Canadian Association of Geographers (Waterloo, 1971).
See also [Canada], *Final Report. Royal Commission on Canada's Economic Prospects.*

These objectives cannot be attained without consistent action by the various governments, as well as by private enterprise.

Personal Expenditures

The Canadian Gross National Product ($92 billion in 1971) comprises four major sectors: personal expenditures, gross business-capital formation, expenditures on goods and services by governments, and the balance between exports and imports. In 1965, out of a GNP of $52 billion, the first two categories accounted respectively for 61 per cent and 20 per cent.

Changes during the Last Thirty Years

Table 14.4 shows the main items of private expenditure for three years: 1939 (prewar), 1964 (prosperity) and 1969 (inflation). The change in total volume of such expenditure is the first striking feature. In terms of a constant dollar, personal expenditures rose from $9 billion in 1939 to $31 billion twenty-five years later. There were equally significant changes in the various components of expenditure. Towards the end of the depression, savings and taxes represented only 5 per cent of total personal income; about 90 per cent of the total went on personal expenses (food, accommodation, household expenses, clothing). Society was more concerned with survival than with growth. By contrast, in 1964 the proportion of expenditure devoted to taxes and savings was almost three times as great (14 per cent). Similarly, expenditure on direct consumption showed a relative decline as compared to increases in spending on transport (due to the automobile and to suburban growth) and on medical care; economically, it is a picture of the affluent society. However, due to inflation, housing and food are by far the major items of consumer spending. In recent years, government medicare schemes have reduced personal expenditure on health services.

TABLE 14.4 Personal Expenditures in Canada
(in millions of dollars)

Purpose	*1939*	*1964*	*1969*
Food	960	6,693	7,634
Housing	582	4,564	8,376
Transport	370	3,656	5,496
Household expenses	593	3,548	3,311
Taxes	62	2,930	7,469
Clothing	479	2,807	3,579
Personal and medical expenses	261	2,563	1,743 (+2,341 to plan)
Net savings	194	2,181	3,388
Tobacco and liquor	228	1,909	2,715
Total expenditure	4,290	35,019	46,535

SOURCE: *Canada Year Book* for 1958, 1966 and 1971.

Personal income is not the only source of the high level of expenditure; to it must be added the effect of consumer credit, which has increased greatly since 1946. This growth began in 1930 through the creation of American subsidiaries, which at present finance approximately 60 per cent of the credit held by Canadian consumers. The advance or the guarantee of money enables the purchase of cars, houses and household goods, among other things, or the refinancing of debts.

Tourism

Of the two types of tourism, international and intranational, we know rather more about the former than about the latter. Despite studies of its historical development, structure, forms and revenues, and of specific tourist regions in provinces such as Ontario, Quebec, British Columbia and Nova Scotia, we still do not know the overall Canadian situation. There is no reliable basis on which we can compare the total tourist expenditures of Canadians within Canada, those by Canadians abroad and those by foreigners in Canada.

What is certain, however, is that tourism constitutes one of the chief sources of external expenditure and of income. By 1964, both were running at about $750 million; by now they have reached the billion-dollar mark. The international side of Canadian tourism is mainly a Canadian-American affair. In 1964, each country received approximately the same number of visitors from the other. In recent years, and particularly during Expo year (1967), Canada's usual tourism deficit of about $100 million has become a favourable balance. In 1970, there were 24 million tourists from the U.S.A. among a total of 37 million entries into Canada. Studies of the reasons for travel by Canadians in the United States have shown that in 1963, pleasure accounted for 47 per cent, visits to parents and friends 34 per cent and business 12 per cent. The importance of these reasons varies according to province, with each of the above being of most importance in Ontario, the Maritimes, and Alberta respectively, a reflection of their different historical and economic links with the U.S.A.

The exchange of tourists between Canada and the rest of the world is a different matter.

Its annual volume is small, scarcely more than half a million people, and there is a substantial deficit.

The Balance of Payments

Two different economic indicators must be clearly distinguished from each other, even though they are usually referred to in similar terms. The (international) *balance of payments* is composed of the balance between "current accounts" and the "capital market". The latter includes various transfers and represents the inflow and outflow of money, in the long term or short term, by private business or the public sector. As for the *balance of current accounts* (or current transactions), it consists mainly of the external balance of trade in goods (exports and imports), tourist transfers, transport costs and especially the servicing of working capital (interest and dividends).

The balance of payments can therefore be considered as an overall balance composed of several specific balance sheets.

The General Balance

The multifarious exchanges of goods and services between Canada and the outside world represents a very complex situation, in which a deficit on current account is financed in part by an influx of foreign capital (see table 14.5).[3]

Current accounts from year to year show both surpluses and deficits. On the one hand, the balance is governed by variations in the quantity of goods exported and imported. Thus, positive trade balances from 1916 to 1926 gave way to deficits from 1927 to 1934, after which there were surpluses again until 1950. For more than a decade, a negative balance in current account had been related mainly to a combination of individual deficits. For example, in 1965, servicing of foreign capital amounted to $700 million; other types of current account, $375 million; cost of transport, $80 million; tourism, $49 million. Between 1962 and 1965, these various deficits, together with a growth in imports related to national prosperity, prevented the balance of current accounts from showing much of the beneficial effect being achieved by a high level of exports. By 1970, the trade balance was once again in surplus.

The difference between the balance of current accounts and that of the capital account is accounted for by changes in gold reserves and international financial arrangements.

Balance of Payments between Canada and the United States

During the first decades of the twentieth century, Canada maintained its balance of payments mainly through a triangular arrangement, in which the surplus on account with Great Britain was used to balance a deficit with the United States. Since then, solutions have become more complicated. Of all the links which bind Canada to its powerful neighbour, those in the economic sphere are not only the easiest to measure but are also the strongest. Table 14.6, for example, shows the positive or negative balances, in 1964 and 1969, in the principal sectors.

Hence, the periodical surplus of imports over exports and the servicing of American capital invested in Canada are the two chief reasons for the deficit on current transactions. In the context of capital movements themselves, the main inflow is related to new Canadian stock issues, which are bought partly by nonresidents.

The perennial problem of the balance of payments will not be solved on a sectoral basis. Since the components of national economic life are extremely interrelated, any solution for the four main elements of the balance of payments —exports, imports, servicing of foreign capital,

3. See M. K. Inman and F. R. Anton, *Economics in a Canadian Setting*, part IV.

TABLE 14.5 Canadian Balance of Payments
(in millions of dollars)

Items	*1956*	*1961*	*1965*	*1970*
Trade balance	− 728	+ 173	+ 118	+3,002
Other current activities	− 638	−1,101	−1,248	−1,705
Total of current accounts	−1,366	− 928	−1,130	+1,297
Long-term capital flows	+1,424	+ 930	+ 864	+ 814
Short-term capital flows	− 10	+ 290	+ 425	− 581
Change in reserves	+1,414	+1,220	+1,289	+ 233
Net capital movement	+ 48	+ 292	+ 159	+1,530

SOURCE: Statistics Canada.

TABLE 14.6 Canada–U.S.A. Balance of Payments, 1964 and 1969
(in millions of dollars; positive and negative with respect to Canada)

	1964			*1969*		
	(−)	(+)		(−)	(+)	
Trade balance	811				367	
Tourism		109			68	
Gold		145			108	
Interest and dividends	652			873		
Transport	97			87		
Miscellaneous	349			316		
Total	1,909	254		1,276	543	
Balance on current account			−1,655			− 733
New Canadian stock issues		1,067			1,497	
Other transfers		554		761		
New direct investment		134			355	
Total		1,755		761	1,752	
Balance on capital account			+1,755			+1,092

SOURCE: *Canada Year Book* for 1966 and 1971.

and new investments—must be based on a comprehensive policy.

At a still higher level of the macroeconomy, a balance must be sought which will take into account, among other things, salaries and wages, corporate profits, costs of production and distribution, prices and the state of the national market.

Finally, because of the special situation in Canada, better economic harmony must be sought at the regional level.

Chapter 15

Canadian Output

Elsewhere in this book, explanations are provided for the quantitative data collected together in this chapter. The values of output, together with the size of the labour force in different industries, are used as indicators in this analysis (see table 15.1). It should be remembered, however, that production of goods is not by any means equivalent to the total sum of all economic activities. The analysis relates only to about 40 per cent of the labour force and 50 per cent of the Gross National Product, since the production of services has not been taken into account.

In terms both of turnover and of number of workers, Canadian production of goods is now dominated by manufacturing and processing. World War II and the postwar period forced Canada into a highly industrialized world characterized by powerful corporations and unions, by a high level of technology, by mass production and by rapid urbanization. Even without adding construction to the value of manufacturing output, Canada no longer appears as primarily a producer of raw materials; it has achieved a significant rise in the industrial hierarchy. From 1870 to 1956, while the value of raw material production increased 2,900 times, that of secondary products (manufacturing and construction) rose 4,000 times. However, this must not be taken to imply that the manufacturing sector has yet reached full maturity.

Because of changes in productivity and because of the nature of the commodities involved,

TABLE 15.1 Output and Labour Force (%) in Various Economic Sectors

Categories	*1962 Value Added*	*1961 Labour Force*
Manufacturing	54.7	47
Construction	17.7	15
Agriculture	11.4	29
Mining	8.2	2
Electric power generation	4.1	3
Forestry, fishing and hunting	3.9	4

SOURCE: *Canada Year Book, 1966* and *Canada 1867–1967.*

there are wide differences between the proportion of workers and the net value of output in the two main sectors of primary production, agriculture and mining. Between 1946 and 1964, the average rate of growth of the mineral extraction industry was 8.8 per cent, that of agriculture only 1.5 per cent (compared to 4.1 per cent for the GNP). It seems inevitable that mining, in terms of value of production, became more important than agriculture; also, in 1968, the value added in mining had a percentage slightly superior to that in agriculture.

The Manufacturing Sector

Four Stages of Industrial Development

(1) The first stage was necessarily domestic in character, concerned with the production of articles which were difficult to import. Thus, even though the French regime certainly did not encourage manufacturing, wheat mills, forges, and workshops producing clothing, beer and wood products gradually grew in number.

(2) During the first century of British rule, the economy benefited from the experience of the Industrial Revolution in Britain. In Great Britain it also found capital, industrial leaders and markets. Canada became an important producer of wooden ships, lumber, potassium and, even at this stage, wheat. Around 1870, the group of industries which depended almost entirely on the output from agriculture, fishing and forestry represented almost half the total value of Canadian manufacturing.

(3) From 1870 to about 1910, there was a transitional period characterized by railway development (which altered the system of supply and the nature of the market), by the protection policies of 1879 (the industrial effects of which were small), and by the growth of the Canadian market due to immigration and the new needs of the Prairies. The nature of Canadian industry also changed; around 1890, paper production and food industries became relatively more important than industries based on woodworking. This implied a decline in importance for the Maritimes, especially New Brunswick. From 1900 to 1910, metal-working industries assumed a significant place in the economy for the first time. Ontario contributed more than half the manufacturing output, much of which was destined for the Canadian market.

(4) The last stage was one of industrial power and increasing industrial maturity. In the twentieth century, a variety of factors—the large-scale development of hydroelectric power; mineral and forest resources; the use of technology, both in the production and the conversion of raw materials; the growth of the national market; financial and management assistance from the U.S.A.; new industrial oases in the Middle North; Commonwealth preferences (especially for cars); and even international rivalry—have all combined to ensure the development of industry, representing a value added in 1968 of around $18 billion.

During this fourth period, the sequence of prosperity, depressions and recessions has affected the stability of Canada's major industries in various ways. Lumber mills, the car industry and steel were severely hit by the depression (see table 15.2 for 1933). During the war, controls on output and export led to abattoirs and canning factories taking first place and kept metal refining (including nickel, aluminum and copper) in second place. The war caused a revival in steel, the automobile industry and electric accessories, but forced a decline in paper, dairy products and flour milling. The emergence and growth of foreign markets as well as strong domestic postwar demands gave renewed vitality to the older wood industries (even the sawmills), flour milling, dairy

Table 15.2 Order of Importance of Manufacturing Industries by Gross Value of Output, 1911–1968

Categories	*1911*	*1922*	*1933*	*1944*	*1946*	*1958*	*1968*
Pulp and paper	6	2	1	5	1	1	2
Petroleum	8	9	6	14	8	2	4
Nonferrous metals	5	−40*	2	2	3	3	10
Meat and food processing	2	3	3	1	2	4	3
Vehicles	9	6	11	7	9	5	1
Basic iron and steel	4	20	31	13	13	6	5
Sawmills	10	4	14	11	4	7	8
Butter and cheese	3	5	5	10	6	8	7
Electrical products	7	17	16	8	7	10	−40*
Milling	1	1	4	12	5	32	−40*

* Not among the first forty industries.

Source: 1911 Census, and *Canada Year Book* for 1951, 1961 and 1971.

products, petroleum and electrical products and kept meat production and metal refining important. A few years ago, the discovery of iron and petroleum reserves led to growth in related industries, while the recession of 1958 affected consumer goods (electrical accessories). Meanwhile, primary industries (abattoirs, lumber mills, dairy products) gave way to more highly developed manufacturing industries (e.g., automobiles). For several decades the two most stable industrial groups have been pulp and paper, followed by refining; these, it should be noted are two very important secondary industries based on export. Importance, industrial development, capital and foreign markets are closely interlinked. For the last few years, the stability of these two main Canadian industries has been less sure. Petroleum and vehicles are moving to a position of primacy.

Manufacturing

Due to the introduction of new methods of compiling industrial statistics after 1960, the comparison of the data with those of preceding decades is difficult. From table 15.3 it can be seen that Canadian manufacturing industry is oriented, to a considerable extent, towards production goods. This emphasis on durables reflects the enormous mining and forestry resources, the needs for transportation, the demands of external trade and the relatively limited national demand for perishable goods. In terms of employment some categories, such as the food, clothing, textile and wood industries, are good generators of jobs; others provide relatively few jobs, for example the chemical industries. In terms of income, the transport, metal working and paper industries pay higher wages than textile manufacturing.

Depending on the nature of raw materials, the elasticity of demand for the product, market location and the proximity of the United States (this last is important in the location of branch plants of U.S. firms in Canada), some industries are highly concentrated, including oil and gas, primary metals, and even paper making. Other industries are widely scattered, such as food industries, lumber, publishing and clothing manufacture.

TABLE 15.3 Leading Manufactures by Industry Group, 1968

Group (Industry)	*Employees*		*Salaries and Wages*		*Value Added*	
	Number (in thousands)	*Rank*	*Dollars (in millions)*	*Rank*	*Dollars (in millions)*	*Rank*
Food and drink	141	1	668	2	2,636	1
Transportation equipment	108	2	747	1	2,045	2
Primary metals	86	5	570	5	1,514	3
Metal products (excluding machines and transport)	101	3	587	3	1,493	4
Paper and related products	88	4	584	4	1,479	5
Chemicals	38	12	232	12	1,285	6
Electrical equipment	76	8	401	6	1,173	7
Printing and publishing	48	10	289	9	916	8
Forest products	77	7	396	7	896	9
Machinery (excluding electrical)	44	11	269	10	737	10
Nonmetallic minerals	37	13	223	13	686	11
Textiles	57	9	253	11	652	12
Clothing	85	6	364	8	600	13
Total	1,160		6,278		18,332	

SOURCE: *Canada Year Book, 1971.*

Paper

This industry's significance is due to its historical role, its absolute quantitative importance and to the fact that it displays many Canadian industrial characteristics. Closely associated with that development of manufacturing at the beginning of the twentieth century which placed renewed emphasis on the production of staples for export to the United States, the paper industry has developed to the point where Canada is among the world leaders, along with the U.S.A., Japan, the U.S.S.R. and the Scandinavian countries (see table 15.4). Its establishment was dependent on the abundant forestry resources (spruce, balsam, fir, hemlock and even deciduous species) and hydroelectricity, on a ban on pulp exports, on a convenient drainage pattern and on available labour. The wide variations in levels of demand, the similar nature of the production process and the great amounts of capital required led to the creation of large private corporations.

The industry consists of four major processes, beginning with the production of pulpwood (18.3 million cords in 1968), 95 per cent (by volume) of this going to Canadian plants. Export of the raw material is virtually nil. The second stage consists of the manufacture of the pulp itself (16.7 million tons), about 60 per cent of which takes place in Quebec and Ontario. Export, mainly to the United States, represents about one-quarter of the total. The third stage is primary paper production (11.1 million tons), 43 per cent of which comes from Quebec alone. The chief product (74 per cent) is newsprint; for a long time, Canada has been the largest producer and exporter of this product. The U.S.A. is the main international customer; during the 1960s, it has taken 83 per cent to 88 per cent of the output, though in 1968 the proportion fell to 74 per cent. Finally, from the primary paper, Canada produces a wide variety of special papers to meet a demanding national market.

The three main problems facing the industry are the determination of the maximum sustained yield of forest resources, comprehensive forest management and the stability of external markets for newsprint. In 1972, the forest industry in Quebec was in a critical state, and was seeking special advantages from the provincial government.[1] British Columbia, though, is becoming a major producer of pulp and paper.

1. See *The Competitive Position of the Quebec Pulp and Paper Industry*, report by the Council of Pulp and Paper Producers of Quebec (Quebec, 1972).

TABLE 15.4 Pulp and Paper Production, 1965
(in millions of tons)

Country	Production
United States	76.9
Canada	24
Japan	13.2
U.S.S.R.	10.1
Finland	9.7
Sweden	9.6
Germany	6.9
China	5.2
France	5.1

SOURCE: *Financial Times*, 1967.

TABLE 15.5 Value Added in Manufacturing, 1963 and 1968 (in thousands of dollars)

Province	*1963*	*1968*
Newfoundland	76,900	88,386
Prince Edward Island	11,300	16,569
Nova Scotia	193,300	261,044
New Brunswick	177,692	240,753
Quebec	3,747,000	4,855,896
Ontario	6,899,300	9,714,889
Manitoba	416,500	443,002
Saskatchewan	144,800	170,002
Alberta	413,800	604,529
British Columbia	1,090,000	1,575,436

SOURCE: *Canada Year Book* for 1966 and 1971.

Regional Aspects

Of all the factors influencing industrial location, four seem to dominate: the location of certain raw materials which are difficult to transport, the proximity of the United States (from where both the initiative and the customers frequently come), the main nodes of the Canadian market, and transport facilities. The influence of United States industry, which has 800 subsidiaries in Ontario, caused D. M. Ray to develop the concept of an "economic shadow", which may account for the industrial differences between eastern Ontario and southwestern Ontario.[2] The latter area, which is better placed to benefit from the vitality of the industrial heartland of the United States, has a more highly developed industrial structure than the more peripheral area of eastern Ontario. The four factors together have created a tendency for industry to establish itself in southern Canada and especially along the axis of the lower Great Lakes and St. Lawrence (along the lakeshore in Ontario and in southwestern Quebec).[3] These areas accounted for more than 80 per cent of the value of Canadian industry. Within these two main industrial ecumenes (both of them in eastern Canada), a variety of factors, such as available land, canals (e.g., the St. Lawrence Seaway), railways, port facilities and a number of pre-existing locations (e.g., steel making at Hamilton) have determined the precise sites of different activities.

The Dominance of Ontario

The industrial power of Ontario is long-established; as early as 1880 the province already accounted for 51 per cent of the Gross National Product of Canada. Though industrially more diversified than any other province, it is among the best-developed areas in regard to most products, including automobiles, electrical appliances and machinery. With a few exceptions (e.g., paper making is only the

2. D. M. Ray, "The Location of United States Subsidiaries in Southern Ontario", in R. L. Gentilcore, ed., *Canada's Changing Geography*, pp. 149–162.

3. See [Quebec], *Géographie de l'industrie manufacturière du Québec*, Vols. 1 and 2, studies by the Ministry of Industry and Commerce (Quebec, 1970).

seventh industry in Ontario), the order of importance of the chief Canadian industries and the order in Ontario resemble each other. Industry in Ontario is less centralized than in Quebec, British Columbia or Manitoba. Half the main Canadian industrial centres are located in Ontario.

Canadian industry is strongly oriented towards the production on the one hand of partly-finished raw materials for export and on the other of a fairly complete range of industrial products for the home market. The next stage in development, with more dangers but seeming to fit the international character of the Canadian economy, would involve mass production of more specialized products for a competitive world market. By such changes to its industrial structure, Canada would truly become an industrialized nation.

The Primary Sector

It should be kept in mind that the statistical distinction between primary activity and the manufacturing sector is not always very clear.

Commercial Agriculture: A "National" Enterprise

Agriculture has a long history. Archaeological pottery remains suggest that it was practised by the Iroquois; the method of boiling down maple syrup was certainly borrowed from the Indians. Since the arrival of Europeans in Canada, there have been four main agricultural periods.

(1) *Before 1780*. At the beginning of the seventeenth century, a few French and British settlements were established in Newfoundland, in Acadia and on the banks of the St. Lawrence. Except in the marshes of the Bay of Fundy, techniques were simple. Because output was insufficient, imports from the mother countries and from the West Indies remained essential.

(2) *From 1780 to 1880*. To some extent, thanks to the Loyalists, the ecumene grew and the centre of agriculture was displaced westwards into Ontario. Agriculture became more organized with laws, government services, agricultural societies, schools and regional markets. It also became commercial.

(3) *1880 to 1926*. This period was the most important, first because of the extraordinary growth of the farming area; this area expanded from 36 million acres (around 15 million hectares) in 1871 to 163 million acres in 1931 (65 million hectares). Within a few decades, in other words, an area the size of France had been utilized for agriculture. Some 30 per cent of this area was at first granted to the railways and to the Hudson's Bay Company. The centre of gravity of Canadian agriculture was displaced westwards to the Prairies. This massive colonization could not have taken place without the railway, the Homestead Act of 1872, the arrival of millions of immigrants and the development of international markets. Around 1910, wheat and flour became the chief Canadian items of export.

(4) *Since 1927*. During the depression, despite a "back to the land" movement, the agricultural area scarcely extended and incomes decreased sharply. The government tried to minimize soil and marketing problems, but western agriculture became a national issue. The crisis led to the establishment of the Canadian Wheat Board in 1935 and to the Prairie Farms Rehabilitation Act. At the same time, a cooperative market and farm credit were developed. World War II and recent massive sales have revived cereal cultivation. However, even though wheat is a main cash crop, wheat dominance is at an end.

Elsewhere in Canada, mixed farming developed in two directions: the dairy industry (cheese and butter, as early as 1860) and specialized cultivation (tobacco, fruits). In recent decades, agriculture throughout Canada has tended to become more rational and more scientific. Its output has grown despite a large decline in the agricultural labour force and a reduction in the area under cultivation.

The Agricultural Ecumene

The two chief foci of agriculture are, on the one hand, Saskatchewan and Alberta (about two-thirds of the Canadian total) and, on the other, the three Shield provinces: Manitoba, Ontario and Quebec (around 30 per cent). The Prairies as a whole represent three-quarters of the area of agricultural Canada. Quebec, despite its popular image, has a cultivated area only the size of that in Denmark.[4] In terms of commercial output and crop systems, Canada may be divided into three major regions, each containing a number of dispersed cells:

(1) East of the Frontenac axis in eastern Ontario, agriculture tends to be old-fashioned and often marginal, apart from a few specialized oases of stock raising, vegetables and fruits.
(2) Southern Ontario, the Prairies and some valleys of British Columbia are areas of commercial cultivation.
(3) On the northern fringe of southern Canada, there still exists a pioneering agricultural margin with little development, e.g., the Pasquia Land Settlement Project on the Manitoba-Saskatchewan border.

The average size of farms varies from province to province. At the extremities of Canada, in Newfoundland and British Columbia where agricultural opportunities are limited, farms are usually from 10 to 69 acres (4 to 30 hectares). On the Prairies, in contrast, 54 per cent of the farms have an area of more than 400 acres, since mechanization and the type of cultivation favour large-scale activity. The soil itself can no longer be regarded solely in pedological terms. The land is a complex phenomenon, the product of errors such as overintensive use, of problems such as weeds, and of management such as manuring, the use of fertilizers, irrigation, drainage, fallowing and rotation. Experts believe that the Canadian agricultural ecumene could grow by another 40 million acres (16 million hectares) for cultivation and by another 55 million acres (22 million hectares) for stock raising.

In regard to soil erosion, the Agricultural Institute of Canada distinguishes between humid eastern Canada, where only 4 per cent of the cultivated soils suffer from severe erosion (defined as a one-third reduction in productivity) and some of the grain-growing areas of the Prairies where this erosion reaches 25 per cent. The recent practice of harrowing tends to reduce soil erosion.

Income

Table 15.6 shows the monetary income from some of the main crops in 1969. The total income for the whole of Canada was $4.195 billion.

The attention usually devoted to wheat makes the significance of the total value of all animal products, including dairy products, all the more surprising. The enormous production of meat, the basis, as we have seen, of major manufacturing industries, reflects the high level of national consumption. The importance of stock rearing puts Ontario at the head of the list in terms of agricultural receipts. It should also be noted that despite the reputation of the Canadian apple ($40 million in 1968), fruit cultivation ranks a very poor third.

Wheat, the Agricultural King

In terms of land under cultivation, quantity produced and value, wheat clearly dominates

4. See H. Morrissette, *Les conditions du développement agricole au Québec* (Quebec: Laval University Press, 1972).

TABLE 15.6 Cash Receipts from Farming Operations, 1964 and 1969 (in thousands of dollars)

	1964	*1969*
Crops		
(1) Wheat	741,000	697,172
(2) Tobacco	96,000	144,941
(3) Barley	72,000	95,265
(4) Potatoes	55,000	67,034
(5) Oats	33,000	32,500
All crops	1,503,000	1,392,126
Livestock:		
(1) Beef and veal	645,000	969,589
(2) Dairy products	531,000	763,930
(3) Pork	325,000	461,303
(4) Poultry and eggs	305,000	438,091
All livestock	1,848,500	2,689,921
Tree crops:		
(1) Fruit	71,000	87,374
(2) Maple syrup	5,000	18,259
All tree crops	103,000	105,633

SOURCE: *Canada Year Book* for 1966 and 1971.

the major crops. Apart from Ontario, which grows more winter wheat, Canada produces mainly spring wheat. This is certainly the case on the Prairies, especially in Saskatchewan, which from 1955 to 1959 accounted for 63 per cent of the national output. As its automobile licence plates proclaim, it is very much a "wheat province".

The problems of wheat cultivation begin with the annual variations in output (table 15.7), resulting in large variations in national income and in export receipts; the poor years bring problems because the capital invested does not decline in the same proportion.

Canada is a producer of wheat for export (table 15.8), the ratio of external sales to yearly output varying from 60 per cent in 1949 to 85 per cent in 1958 and to 50 per cent in 1970. In 1964, international sales of Canadian wheat reached about 600 million bushels, more than half of which went to socialist countries. In 1972–73, 144 million bushels are being sent to China. Wheat buyers change with the times. The future market in western Europe will depend to some extent on the new tariff regulations.

Agriculture, with a capital investment of more than $10 billion and a major contributor to GNP and to transport activities, national consumption, domestic industry and exports, remains one of the bases of the Canadian economy. In a balanced economy, however, Canada should define the minimum role required by agriculture in an urbanized society

TABLE 15.7 Canadian Wheat Output
(in millions of bushels)

Year	Output
1908–1940 (average)	310
1942	556
1943	284
1952	688
1954	331
1966	844
1969	684

TABLE 15.8 Canadian Wheat Exports
(in % of sales)

Country of Import	*Average 1935–1939*	*Average 1960–1964*	*1969*
U.K.	62	24	18
U.S.A.	21	16	4
Others	17	60	78

SOURCE: *Canada Year Book, 1971.*

TABLE 15.9 Canadian Wheat Balances
(in millions of bushels)

	1948–1949	*1957–1958*	*1963–1964*	*1969–1970*
Production:				
(1) In storage on 1 August	77.7	729.5	487.2	851.8
(2) During crop year	386.3	370.5	723.4	684.3
(3) Imports	0.3	0.08	0.05	
Total	464.3	1,101.1	1,210.7	1,536.1
Consumption:				
(1) Exports	232.3	315.5	594.5	346.5
(2) Canadian consumption	129.6	169.6	156.7	178.6
Total	361.9	485.1	751.3	525.1
In storage 31 July	102.4	615.8	459.4	1,011.0

SOURCE: *Canada Year Book* for 1950, 1959 and 1971.

and then redevelop abandoned agricultural land with a wider economic vision and in better accord with regional economics. This could not be done without careful planning by the farmers themselves. Besides, it can be argued that the huge needs of the Third World must surely lead to a human if not a financial solution to the problem of permanent wheat overproduction. In mid-1970, the carryover amounted to more than a billion bushels.

Mining

Mining has an importance which is much greater than is implied by its rank among primary activities (see table 15.10). Manufacturing industry utilizes twice as much raw material from mining as from agriculture and exports of industrial minerals are larger than sales of wheat abroad. Mining in Canada has a long history: in 1578, the explorer Martin Frobisher hoped to extract gold from Kodlunarn Island in Warwick Fiord, Baffin Island. Mining production accounted for less than 2 per cent of the GNP from 1870 to 1890, but now provides nearly 7 per cent. Canada takes fifth place in the world in terms of total value of mineral production (more than $4.7 billion in 1968) and is either first or second in the production of eight individual minerals.

Categories of Mining Products

Major development has taken place both in metallic minerals and in fuels. As late as 1920, mainly due to Appalachian coal (which alone represented 36.2 per cent of the total value of Canadian mineral production), fuels were of greatest importance. A few years later, however, gold, copper and nickel from the Shield caused metallic minerals to rise to the top of the table; by 1937 the latter accounted for 73 per cent of the Canadian total. At that time, the mining industry was very specialized. Since then, the two main groups have altered their relative position to a certain extent. Fuels have gained more from oil and gas development than metallic minerals have gained from iron-ore extraction. Expressed another way, growth has been more rapid for fuels, which, in terms of volume of production, went from a base of 100 in 1949 to 554 in 1964; during the same period, metals grew only from 100 to 210. However, from 1966 to 1970, all metallic minerals grew more quickly than fuels.

This development has also led to changes in the relative importance of the main products. By 1966, oil (19.8 per cent) was first in importance, iron was third and gas sixth. In contrast, coal had fallen to fourteenth place and gold, which in the period 1940–1950 was at the top of the list, had slipped to tenth position. By 1970, the four leading individual products were crude oil, nickel, copper and iron.

Iron. Apart from some iron produced for casting, as at Saint Maurice from 1737 to about 1888, Canada produced little iron before 1923,

TABLE 15.10 Canadian Mining Output
(as a percentage of total value in each year)

	1896	*1926*	*1946*	*1956*	*1966*	*1970*
Metallic minerals	35	47	57.7	55	49.8	54
Nonmetallic minerals	5	6	8.7	7.7	9.3	8.5
Fuels	38	28	20.5	24.9	29.1	30
Construction materials	19	16	13.1	12.4	11.8	7.5

SOURCE: *Canada Year Book, 1966* and [Statistics Canada], *Canada 1972.*

when all activity ceased. Nova Scotia imported from the colony of Newfoundland while Ontario took its supplies from Minnesota. During World War II, however, iron began to be mined again near the Great Lakes (Michipicoten in 1939 and Steep Rock in 1944). Newfoundland entered Confederation in 1949 and Quebec-Labrador began production in 1954. The volume has increased enormously, though there have been wide variations from year to year (see table 15.11).

In 1966, thanks to Labrador, Newfoundland (40 per cent) and Quebec (34 per cent) produced 74 per cent of the total volume; Ontario (20 per cent) and British Columbia (5.3 per cent) were well behind. On the island of Newfoundland, the Wabana mines closed after seventy-seven years of activity.

Characteristic of recent years has been the use, after concentration, of low-grade ores. The annual capacity of the mines at Carol and Wabush near Labrador City reached 16 million tons in 1967. Transport costs are one of the determining factors in the mineral movements. From Schefferville to Pittsburgh, it costs $7.76 per ton via the Atlantic, but $8.30 by the St. Lawrence Seaway. The interests of the steel companies, return freight, and long-term contracts explain why in 1962, more iron left the north shore of the Gulf of St. Lawrence by the Atlantic coast (9 million tons) than by the St. Lawrence route (6 million). Up to the present,

TABLE 15.11 Canadian Iron-Ore Output
(in millions of tons)

Year	*Output*	*Year*	*Output*
1915	0.39	1958	15.70
1934	0	1959	24.40
1939	0.11	1961	20.30
1949	3.20	1966	36.20
1953	5.80	1970	53.00
1956	20.10		

SOURCE: *Canada Year Book, 1966.*

TABLE 15.12 Mining Output by Major Regions
(as a percentage of the value in each year)

	1937	*1966*	*1970*
Ontario	50.3	24.1	28.3
British Columbia	15.1	7.9	8.6
Quebec	14.2	19.2	13.8
Prairies	12.4	35.6	36.7
Atlantic Provinces	7.2	10.1	9.0
Territories	0.8	3.1	3.6

SOURCE: Department of Energy, Mines and Resources.

Plate 15.1 Bulk carriers loading iron ore at Sept-Iles, Que. The rail tracks are designed to allow rapid unloading of the ore into stockpiles, from which conveyor systems lead directly to the waiting ships.

steel works in Ontario receive from Quebec-Labrador only part of the iron ore they use.

Since Canada itself only uses 10 million tons of ore per year, the country has become one of the most important exporters of this primary material. From 70 to 80 per cent of the exports go to the United States, the rest going to western Europe and Japan.

Table 15.12 shows the rank in mining of Canadian megaprovinces. Despite fairly strong variations from one period to another, Ontario's relative importance, related to metal mining in the Shield, has greatly declined as a result of the growth of fuel production on the Prairies. Alberta alone, with almost 22 per cent of the Canadian total in 1966, is becoming the leading Canadian mining province; iron benefited Quebec and Labrador to a much lesser extent. It should be noted finally that the Shield areas are still the source of the greater part of Canada's mineral output.[5]

Mining output, characteristic of the Canadian economy as a whole, with its large total and small amount of processing, is mainly oriented

5. See [Canada], *Principal Mineral Areas of Canada* (Ottawa, 1972), map 900A.

to export markets; these account for three-quarters of the total value. In 1963, 95 per cent of the nickel was exported, 94.8 per cent of the asbestos, 88 per cent of the iron, and 67 per cent of the copper. These figures are quite representative of the present situation. However, two-thirds by value of the production of petroleum and of gas is used to satisfy national needs. Canada would now like to sell more of its fuel to the United States, since it produces more liquid hydrocarbons than it consumes (1971).

Forest Industries

Forest exploitation has always been important in the development of southern Canada. In the east, its economic history has involved four main products: the old industry of potash (derived from wood ash), square timber (during and after the Napoleonic blockade in Europe), sawn timber (which benefited from steam power and from the markets in the western United States) and, in the twentieth century, the major pulp and paper industry. British Columbia, which alone contains half the forest volume of Canada, has gone through five stages of development, according to W. G. Hardwick:

(1) a pioneering era, from 1860 to 1884
(2) a speculative phase, from 1885 to 1908
(3) a period of growth, from 1909 to 1929
(4) the depression, followed by wartime revival
(5) since 1946, a period of sustained growth[6]

The Vegetation Cover

Most of the terrain of Canada is covered by vegetation, part of which is not visible during the snow season.

Factors affecting vegetation. The cover types are a function of four sets of factors: intrinsic fertility of the soil, amount of incoming heat energy, water content and man's activities. Present-day plant distributions are also a function of past conditions, including the disappearance (or preservation in refuges) of vegetation during glacial periods, recolonization during the interglacials, and northward or southward migrations in response to postglacial temperature and moisture changes.

Vegetation types in the ecumene. There are several available methods of classifying plant covers. In southern Canada, however, a variety of criteria enables the recognition of a number of broad zones, which may nevertheless contain within them major local variations.

In eastern Canada, two *ecosystems* can be distinguished (see fig. 1.3): firstly, the mixed forest of the Great Lakes–St. Lawrence between Lake Superior and Newfoundland, where maple forest (*Aceretum sacchari*) is the characteristic component of the climax stage; and secondly, farther north (as far as 52°N), the commercial part of the enormous concave zone of the boreal forest or, as it has been more appropriately called, the temperate coniferous forest.[7]

In western Canada, two other vegetation zones are found in the settled ecumene, or close to it. On the west coast, the Pacific forest or coastal forest, which extends from California to the Gulf of Alaska, is one of the most important coniferous forests in the world. At Britannia Beach, an average of 76 inches (193 centimetres) of precipitation falls per year and the temperature exceeds 42°F (5.6°C), the threshold for plant growth, for 246 days of the year. The forest itself is huge and the annual incremental growth is high. The main constituents are the

6. See W. C. Hardwick, *Geography of the Forest Industry of Coastal British Columbia*, Occasional Papers, B.C. Division, C.A.G., No. 3 (Vancouver, 1963).

7. J. Rousseau, *Aperçu biogéographique des régions nordiques du Québec* (Quebec: Centre d'Études Nordiques, 1967), p. 10.

western red cedar (*Thuya plicata*), western hemlock, Sitka spruce and many firs, including the Douglas.

The other great vegetation zone of the western ecumene occupies the southern half of the Prairies. It consists of two roughly parallel arcs which are extensions of zones in the United States. The northern crescent consists of a mixture of forest and prairie, the parkland. The prairie, often located on excellent soil, consists of tall grasses, including *Andropogon scoparius*. In contrast, in southeast Alberta and southwest Saskatchewan, there is an area of much shorter grasses (sometimes 6 inches or 15.2 centimetres), or scattered hummocks. The landscape is rather like a steppe; this semiarid section unfortunately became known in early survey reports as *Palliser's Triangle*.[8] Both of these formations have a water deficit during the latter half of summer.

8. J. Warkentin, *The Western Interior of Canada*, p. 232.

Plate 15.2 The MacMillan-Bloedel sawmill at Chemainus, Vancouver Island. Timber is trucked from a twenty-five- to thirty-mile radius and sorted in the water beside the mill. In the foreground, ships load sawn timber, while wood chips are sent by barge to the Harmac pulp mill at Nanaimo. British Columbia's forest resources could yield double their present output on a sustained-yield basis, and still more through the use of more advanced silviculture techniques.

The present landscape of the steppe-prairie is no longer that which existed "before man", or rather before the arrival of nonindigenous man. The colonist, through his cultivation of the soil, through extensive stock rearing, through the erosion that he caused and through the trees he planted, has become an important ecological agent. On the one hand, the tree (the aspen) has followed man in his conquest of the prairies and windbreaks have grown up; in other places the steppe has become even more barren.

A few dozen miles farther north, the boreal forest occurs again, marked by the reappearance of conifers. As in eastern Canada, this forested zone is several hundred miles wide, but the tree density is less, due to the decline in average annual precipitation from the Atlantic to the Rockies. West of Lake Winnipeg there has, as yet, been little exploitation of the boreal forest for paper, though there are numerous small sawmills on the margins of the settled ecumene. However, some huge plants have recently appeared, such as the Churchill Forest Industries near The Pas, Manitoba.

Exploitation

The fact that the chief forests are found beyond the main ecumene explains why farm woodlots form only 3 per cent of productive forests and why the largest forest owners are governments (Crown forests). Concessions for the exploitation of forest resources have, however, been granted to very large private corporations like MacMillan-Bloedel. Methods of exploitation are very advanced, especially in British Columbia, and winches, tractors, loaders and mechanical unloaders have permitted the handling of huge Douglas firs, before which unassisted man would remain powerless. The practice of clear-cutting, still widespread, leaves behind a desolate landscape. Elsewhere, the felled areas crisscross the forested areas.

The respective proportions in the various categories of table 15.13 vary as a function of external markets, national construction activity and climatic conditions (which mainly influence fire losses). Despite the importance of the pulp and paper industry, wood for manufacturing is of greatest importance. In this latter regard, British Columbia produces 44 per cent of the total Canadian volume. Quebec takes second place (25 per cent of the total volume cut) because of its massive output of pulpwood. In view of the recent forestry investment in western Canada (more than a billion dollars in 1967) the output of the western provinces will continue to increase.

The Market

In regard to foreign trade, there is little in the way of imports, while exports of unprocessed wood account for only 3 per cent of the volume of cut timber. Taken as a whole, Canada sells more wood products than it uses itself and they

TABLE 15.13 Commercial Wood Products in Canada, 1957–1966 (mean) (as percentage of total volume)

Product	%
Timber and logs	49.1
Wood pulp	31.4
Firewood	5.9
Other products	1.4
Losses due to fire	12.2

SOURCE: *Canada Year Book, 1971.*

TABLE 15.14 Exports of Wood Products
(in millions of dollars)

Product	*1964*	*1969*
Newsprint	835	1,126
Timber	515	768
Woodpulp	461	753
Plywood	67	80
Other paper and cardboard	58	123
Logs (crude wood)	34	78
Total	1,970	2,927

SOURCE: [Canada], *Canada, One Century, 1867–1967*, and *Canada Year Book, 1971.*

accounted for a quarter of all Canadian exports until 1969, when the proportion dropped to a fifth.

A rational forest policy ought to include at least two objectives: what is cut should be replaced and the cleared land should be managed according to a broad, multisectoral view of resource opportunities. Secondly, the trend towards the total use of felled timber must be accelerated by making use of the bark, the base of the tree, the heart of the wood, waste from sawing up and even the pile of sawdust.

Two Minor but Long-Established Activities

Fish and fur, which once were characteristic of the Canadian economy, now no longer represent significant components of the GNP.

Fish

There are three main types of fishery:

(1) for personal consumption. After having played an important role in the diet of the natives, explorers, coureurs de bois and even of some settlers, this is now of little significance except for small groups of indigenous people.

(2) sport fishing, either through private clubs or individually. This sport is sometimes costly when it involves using private aircraft to go as far as the Arctic in search of arctic char.

(3) commercial fishing.

There are three main areas: the Atlantic coast, the Pacific coast, inland waters. Though the east coast produces twice as much as the west, British Columbia is the most important single province. Each area has its own characteristic products: the Atlantic provides mainly deep-water fish and shellfish, while the Pacific concentrates on estuarine and coastal species. Despite the existence, until recently, of religious restrictions on meat eating, and despite the historical importance of fishing and the increase in the size of the national market in the twentieth century, Canadians do not eat much fish; the country only uses about a third of its own output. As with other primary products, therefore, Canada is an exporter; in terms of volume of sales, it is exceeded only by Japan and Norway. Other countries, like Portugal, the U.S.A. and the U.S.S.R., are using the marine waters peripheral to Canada for offshore fishing. Canada is not happy with this competition.

The Fur Trade

For several centuries, the export of fur was the basis of the Canadian economy.[9] Today, although production counts for almost nothing in terms of national income, it remains a major element in the international fur market: one-quarter of the world's furs from wild animals come from Canada. Canadian production is obtained from two sources:

(1) trapping, which in 1968 accounted for 45 per cent of the total value. By value, the main item is the beaver (half a million skins per year);

(2) fur farming, which has been developed since 1890.

As there are now more than 200 shades of mink, it is inevitable that breeding yields higher incomes than trapping. The majority of farms are in southern Canada and not in the Middle North, the area which, under the Hudson's Bay Company, used to be the chief producing region. The majority of pelts are sold in their undressed state.

9. See E. E. Rich, *The Fur Trade and the Northwest to 1857* (Toronto: McClelland and Stewart, 1967).

Conclusion

Abundant raw materials are among Canada's best assets.[10] Because of them, the economy is markedly oriented towards the production of items which contain much less manufacturing input than is typical of the majority of industrial countries. Despite a tendency towards greater processing of raw materials (for example, in refining), some products, like iron ore, are still shipped mainly in their raw state. By value, Canadian industry makes more use of its minerals than of its agricultural or forestry products and Canada imports only 10 per cent of its needs in these three areas. Since some raw materials are not renewable, a policy of conservation ought to be instituted for the benefit of the regions involved.

10. See J. L. Robinson, *Resources of the Canadian Shield.*

Chapter 16

Transport and Trade

Internal Transport

Canoe routes, the nineteenth-century system of canals and railways between Lake Huron and Montreal, the construction of the Canadian Pacific Railway around 1880, the creation through amalgamation of Canadian National Railways in 1923, the steady growth of car ownership in the twentieth century, the beginnings of airlines around 1930, the recent growth of trucking, pipelines and telecommunications, the St. Lawrence Seaway (1959), the construction of development axes into the Middle North and the Trans-Canada Highway (1962)—these have been the major landmarks in the development of communications within Canada. This summary of the main events does not emphasize sufficiently the really fundamental role played by transport; it has been, quite apart from its commercial and technical significance, a matter of vital political importance to Canada.

Water Transport

Whether by river, lake or ocean, water links have been crucial to national economic development (figs. 16.1 and 16.2).

Three Centuries of the Canoe

One of the most rewarding geographical themes requiring further study is that of the canoe, which created a distinctively Canadian type, the coureur de bois. According to G. L. Nute, the voyageurs (mainly French Canadians and *Bois Brûlés*) were simultaneously explorers, soldiers, and settlers, as well as powerful and cheerful

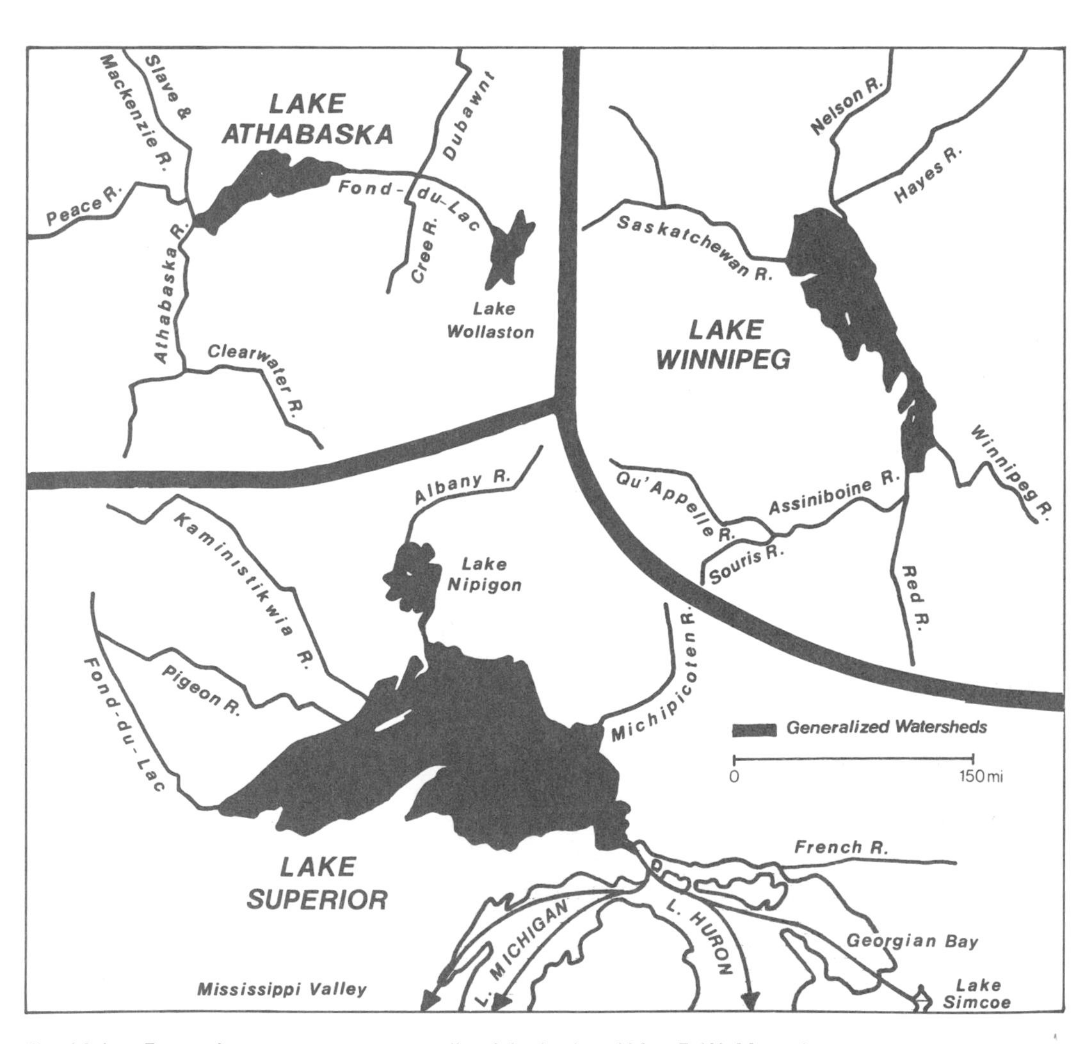

Fig. 16.1 Fur-trade water routes surrounding lake basins. (After E. W. Morse.)

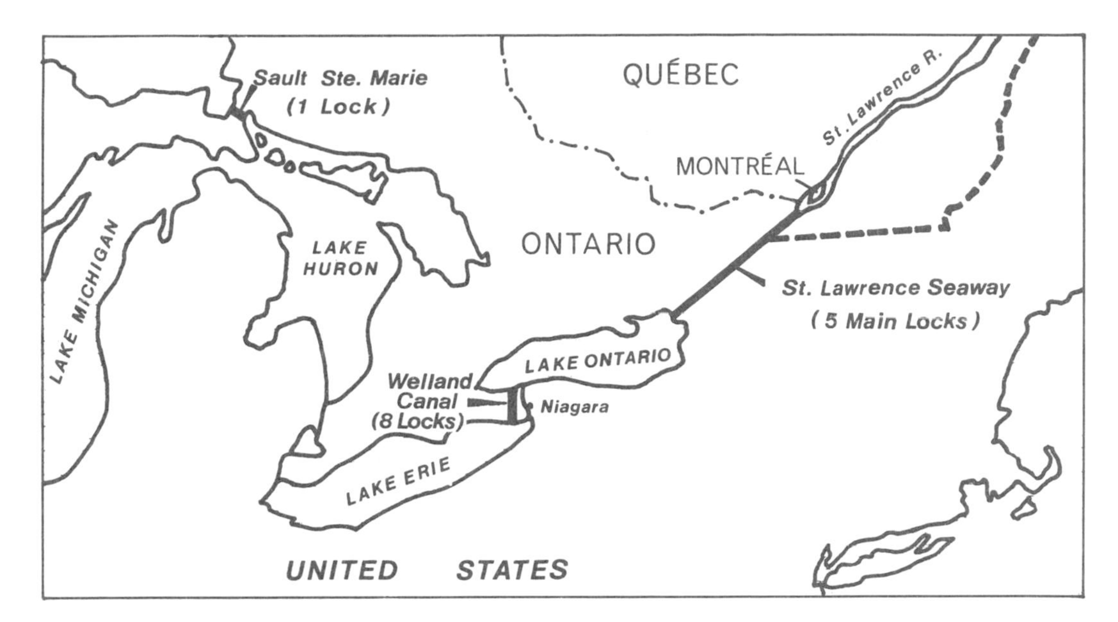

Fig. 16.2 Main canals of Canada.

canoers.[1] The canoe was closely associated with the exploration of the country and with the economic history of the fur trade.

Various forms of craft were developed and adapted to the needs of the users. Before the union of the two main fur-trading companies in 1821, the Scotsmen employed by the Hudson's Bay Company used the York boat, whereas the North-West Company used the famous *canot de Maître*, or Montreal canoe, in the Great Lakes region and the small *canot du Nord*, or North canoe, in the shallow waters of the rivers and lakes of northwest Ontario and the Prairies. After 1763, the "English" merchants of Montreal used the canoe to collect the furs from four main areas: the St. Lawrence, Lake Superior, the Winnipeg region and Lake Athabaska. As it was not possible to make the round trip from Athabasca to Montreal within the five ice-free months which were available, two sets of crews and canoes were used. Each May, these left the two ends of the system, Lachine in the east and Fort Chipewyan in Athabasca. In July or August, they met near Thunder Bay, where the northwestern team exchanged its furs for the supplies brought up from the St. Lawrence. These furs were then moved down to Montreal for shipment before ice closed the river to ocean-going vessels. A similar seasonal cycle governed the activities of the rival Hudson's Bay Company, which assembled the pelts on the southwest shore of the Bay itself. All this had to be done so that the citizens of Europe could have their hats made of felt from beaver skins.

1. G. L. Nute, *The Voyageur*.
See also E. W. Morse, "Voyageurs' Highway. The Geography and Logistics of the Canadian Fur Trade", *Canadian Geographical Journal* LXIII (1961): 141–161; also Vol. LXIII: 2–17 and 64–75.

Canals and Seaways

As was later true of the railways, some canals were developed for noncommercial reasons. An example is the Rideau system which, in the 123 miles between Ottawa and Kingston, has forty-seven locks. Around 1830, it was thought desirable that an inland route assuring continued communication between Upper and Lower Canada should be created, in case the U.S.A. came to control the international section of the St. Lawrence. Other canals were begun to compete with or to imitate the Erie Canal in the United States, south of Lake Ontario. By 1964, Canada maintained close to twenty such routes, the majority of which were around the Great Lakes.

The Canadian canals have two main uses. Firstly, they are a means of transport. Of a total of 97 million tons handled in 1969, the Welland accounted for 53.5 million, the St. Lawrence Seaway below Lake Ontario for 41 million, the Canso Strait in Nova Scotia for 1.2 million and the Canadian locks at Sault Ste. Marie for 1.4 million (whereas the American Soo locks handled 97 million tons).[2]

The Welland Canal, begun in 1829, deepened to 27 feet in 1932, and still being improved, is the most important unit. Eighty per cent of the cargo, most of it in bulk form, is moved by non-ocean-going lakers which have replaced the shallow draft vessels. The St. Lawrence Seaway, authorized in 1951 and opened in 1959, doubled the volume of commodities handled (from 11 million to 22 million tons), but tonnages during the early years have been below the levels anticipated. Even as recently as 1964, 30 per cent of the traffic consisted of iron ore, 18 per cent was wheat, 12 per cent was coal and 3.9 per cent was fuel oil. Five years later, these percentages were: iron ore, 25 per cent; wheat, 9 per cent; coal, 18 per cent and fuel oil, 7 per cent. American interests, which for a long time delayed the building of the seaway, now do not want to see the rise in seaway dues that Canada believes is necessary to repay the capital investment within the planned period of fifty years.

2. See [Canada], *Canada Year Book, 1971.*

In narrow maritime terms, the St. Lawrence Seaway, which can only accommodate medium-sized ships, is of intranational rather than international importance. In 1963, only 5 per cent of the traffic from Lake Superior ports was directly loaded for transatlantic destinations. Deepening the St. Lawrence system has mainly helped the growth of ports in the lower part of the estuary, like Baie-Comeau; these exchange downbound wheat for upbound cargoes of iron ore from Quebec-Labrador. Meanwhile, the St. Lawrence season has been lengthened: since 1885 there has been a gain of seven weeks at Montreal. Technical improvements and a new transport policy could well lead to year-round navigation. In any case, Pierre Camu, president of the St. Lawrence Seaway Authority, anticipates that the seaway will handle 140 million tons by the year 2000, compared to 61 million tons in 1969.

The second main use of the canals is for recreation and sport. Some of them, restricted by numerous locks and shallow draught (less than 10 feet), are very popular with pleasure cruisers. Besides the scenic and historic appeal of these routes, the lock systems are also attractive, as is the fact that there are no tolls. The 123 miles (197.9 kilometres) of the Rideau system, which carried 56,000 small boats in 1966, will probably be comprehensively developed for recreation.

Port Activities

Due to the shape of Canada, its position in the world, the volume of its external and coastal trade (the latter quite as important as long-distance movements), ports have been of great significance in the Canadian economy. Around the middle of the nineteenth century, thanks to shipbuilding, Canada was even a maritime trading nation.

Some ports are highly specialized, such as Thunder Bay and, more especially, Sept-Iles

TABLE 16.1 Tonnage Handled at Major Canadian Ports (in millions of tons)

Port	*1964*	*1968*
Montreal	21.5 (wheat 6, crude petroleum and fuel 6)	17.3 (wheat 2.7)
Vancouver	20.0 (wheat 4, pulp 2)	25.7 (wheat 4.3)
Thunder Bay	18.3 (wheat 10)	13.4 (wheat 5)
Sept-Iles	16.5 (iron 15)	25.9 (iron 24.9)
Port Cartier	10.2 (iron 10)	12.7 (iron 9.6)
Hamilton	9.3 (iron 4, coal 3)	11.9 (iron 5.9)
Halifax	9.1 (crude petroleum and fuel 5)	9.6 (crude petroleum and fuel 6)
Baie-Comeau	8.3 (wheat 4)	4.7 (wheat 2)
Quebec	6.0 (wheat 1)	6.6 (wheat 1.2)
Saint John	5.8 (crude petroleum 2)	5.4 (crude petroleum and fuel 3.3)
Toronto	5.7	5.7 (bituminous coal 2.3)
Sault Ste. Marie	5.6	5.2
Trois-Rivières	4.5	3.0
Sorel	4.3	4.5
New Westminster	4.3	5.0
Total	234.9	245.6

SOURCE: *Canada Year Book* for 1966 and 1971.

and Port Cartier. Others specialize to a lesser extent, such as Baie-Comeau, Hamilton and Saint John, while the largest ports, such as Montreal or Vancouver, and long-established ports, like Quebec, Halifax and St. John's, are the least specialized. Cereals, iron, fuels and wood products are the four chief items handled in terms of volume. All the fifteen largest ports are in eastern Canada, except Vancouver and New Westminster (table 16.1). Equally, reflecting the economic structure of the country, only the two long-established ports of Halifax and Saint John are in the Maritime Provinces.

In the last few years two technological innovations have led to great changes in port activities. Containerization has been responsible for increases in traffic flows in some ports; Quebec City, for example, handled 11 million tons in 1971. Secondly, both the Pacific and Atlantic shores of southern Canada are becoming the sites for "megaports" for intercontinental shipping. Roberts Bank, near Vancouver, is an example.

The importance of the ports listed in table 16.1 should not obscure the vitality of smaller ports which play an essential role in providing services for regions that are not equipped with adequate land transport. These areas include Newfoundland, Labrador, the shores of Hudson Bay and Hudson Strait and the arctic coastline. Tonnage handled is not the only criterion used in evaluating the importance of water transport.

Wheat-Exporting Ports

In his classic study, Henri Baulig described the export of wheat for the year 1926–27 as follows: "48 million hectolitres of Canadian wheat passed through American Atlantic ports,

Plate 16.1 Wharves for wood products and grain at Thunder Bay, Ont., on Lake Superior. Vancouver has now become Canada's leading grain port and shipments from Thunder Bay, during the April to December shipping season, fell from 10 million tons in 1964 to 6 million tons in 1968.

TABLE 16.2 Ports of Export for Canadian Wheat (in percentage)

Ports	*1934–1938*	*1953–1954*	*1961–1962*	*1971–1972*
St. Lawrence, Great Lakes and Atlantic	47	38	47	51
Pacific	28	56	47	45
U.S. ports	23	0.1	0	0
Hudson Bay (Churchill)	2	5	6	4

SOURCE: Calculated from B. Brouillette, *Analyse de courants commerciaux au Canada* (Montreal: Les Presses des Hautes Études Commerciales, 1964), and Canadian Grain Commission, 1972.

but only half of this amount through Canadian ports."[3] Major changes have taken place since that time (see table 16.2). In the first place, prairie wheat moving via the Great Lakes–St. Lawrence barely accounts for half the total shipments. In 1954, the meridian of Saskatoon marked the dividing line for wheat exports via Atlantic or Pacific. A complete diversion of Atlantic wheat exports to Canadian ports also took place before the opening of the seaway in 1959.

Rail Transport

Despite the large marine embayments and a long coastline, Canada is mainly a continental country and real solutions to the main transport problems must be achieved by land-based means.

In the nineteenth century, the expansion of the settled ecumene away from the coastline, islands and peninsulas of eastern Canada depended on land transport. The western end of the Great Lakes represented a real boundary and the Prairies could not be opened up to settlement without a technical breakthrough. Only a century ago, in 1872, an official description by the Department of Public Works in Ottawa of the transport from Fort William to Fort Garry (Winnipeg) was as follows: 45 miles (72.4 kilometres) of cart track to Lake Shebandowan, 310 miles (514.9 kilometres) of waterways and portages ("broken navigation") from there to the Lake of the Woods, then another 95 miles (152.8 kilometres) of cart track. On this long, dangerous and uncomfortable journey, it was necessary to take one's own provisions. If millions of immigrants and easterners were to be attracted to the West, a solution to the problem of adequate transport in northwestern Ontario was essential.

The Railway Period

Three stages can be clearly distinguished.

(1) *Experimentation* began in 1827 at Pictou (Nova Scotia) with the horse-drawn rail transport of coal. In 1836, steam locomotives were introduced between Montreal and Laprairie and these eventually reached Upper Canada.

(2) The *transcontinental phase* began after 1855, thanks to British capital and politics. In relative terms, this was Canada's greatest achievement. Railways became both commercial and political concerns. Although some lines, such as the Grand Trunk around 1860 (from Sarnia to Rivière-du-Loup via Toronto and Montreal) provided the spinal column for a major economic axis and although, forty years later, the route from Skagway to Whitehorse was built in response to the Klondike gold rush, railways

3. H. Baulig, *Amérique septentrionale*, part 1, p. 224.

were built mainly for political reasons.[4] The Intercolonial (from Halifax to the St. Lawrence estuary) was specifically mentioned in the British North America Act of 1867 as a condition of entry of Nova Scotia and New Brunswick into Confederation.

The Canadian Pacific Railway was also a political enterprise in the sense that it was designed to protect the Canadian identity in the West (against the Northern Pacific in the United States) and as a response to the promise made by the federal government to British Columbia in 1870 that it would be linked to eastern Canada. A world depression and the problems facing the project made it very difficult to raise sufficient private and Canadian capital. Only after the "National Policy" was announced in 1878 was a contract signed between the CPR and Canada. Canada undertook to build some of the sections (Port Arthur–Winnipeg), to provide a subsidy of $25 million and also to give 10 million hectares of good prairie land to the CPR. It was from these beginnings that, on 4 July 1886, the first regular train arrived at Port Moody in British Columbia.

Without the extension of railways into the Maritimes and into western Canada, Confederation could not have taken place, at least at the time it did and in the way it did. In the West, the railway certainly had a major economic impact; in fact it was, in 1886, a question "of life or death". There was a close relationship on the Prairies between railways, settlement, cultivation and exports. Later, other companies, reacting against the quasi-monopoly of the Canadian Pacific, constructed other sections of transcontinental routes.

(3) *The era of rationalization.* In Canada, as in the United States, the rail network which developed consisted of a multitude of independent lines with little overall coherence. The ratio of railway mileage to population in Canada was the highest in the world. A commission of inquiry was established in 1916, and seven years later, the government agreed to the amalgamation of the major networks of the Canadian Northern, the National Transcontinental and the Grand Trunk into a vast organization called Canadian National Railways or, more recently, CN. Thus, by necessity, the new nationalized system, the only one to serve all ten provinces and the two territories, began to compete with the privately owned Canadian Pacific, which in many places has a better route location. Taking an even wider view, the amalgamation, or at least the integration, of these two powerful organizations seems desirable.

Recent Changes

In terms of route mileage, trends have been conflicting. On the one hand, the companies have asked the Canadian Transportation Commission for permission to abandon unprofitable lines. On the Prairies alone, although the commission recommended the closing of 8,000 miles (12,874 kilometres) of route mileage, the government in 1966 authorized the closure of only one-quarter of this amount. Within the settled ecumene, the fact that the railways transport only 4.6 per cent of intercity passengers has led to widespread losses. In contrast to this retrenchment, however, approximately 1,200 miles (1,931 kilometres) of new railway have been built into the Middle North since 1946, especially in Quebec-Labrador, in Manitoba and to Great Slave Lake.

By a curious reversal of history, the transcontinental role of the railways is continuously diminishing, and in 1969, the number of trains per week linking Montreal with British Columbia was reduced. Due to a lack of demand, a properly organized transcontinental system has

4. R. I. Wolfe, "Transportation and Politics: the Example of Canada", *Annals of Association of American Geographers* 52 (1962): 176–190.

never really existed, except in the Montreal-Pacific sector. At the present time, using regular schedules, it would take almost two weeks to get from St. John's, Newfoundland, to Dawson in the Yukon, by a combination of railways, ships and buses. Instead of developing a complete and rapid transcontinental system, the railway companies seem to have specialized in intercity services; for example, from Quebec to Montreal and from Montreal to Toronto. The railway is less necessary to the maintenance of Confederation nowadays than it was at the time it was created.

In the face of competition from other means of transport which benefit from investments by the public sector, and because the railways have to some extent become out of step with modern economic conditions, attempts are being made to bring the system up to date. This renovation is, however, far from complete and in areas like Abitibi, there are major demands for further improvement.

New Methods of Transport

World War II was the last occasion when the railway was the dominant form of transport. Since then, major changes have taken place in transport concepts, forms and technology.

Extension of Competition

From 1945 to 1964, the relative proportion of interurban transport carried by rail fell from 72 per cent to 42 per cent, despite a great increase in traffic. Further, despite the seaway and other improvements, transport by water in 1964 accounted for little more than the volume carried by pipelines. Highway transport, which in 1945 amounted to only 5 per cent of that carried by the railway, has grown to 22 per cent. Canada, it has been shown, is in the midst of a period of competition between different means of transport (see table 16.3).

Though rail still dominates freight transport, it is not as popular for passengers. The car, despite its cost and restrictions due to World War II, grew quickly and overwhelmingly (85 per cent in 1964) to become the chief means of intercity transport. The private car has become a normal personal possession; in fact, it is estimated that over most of Canada, "saturation", at 0.45 cars per head of the population, will be reached by 1980.

The Highway System

Highways designed to serve needs other than local ones were slow in coming. It was only in 1734 that the great surveyor Lanouillier de

TABLE 16.3 Intercity Freight Movements in Canada, 1928–1964
(percentage of total volume)

Year	*Rail*	*Water*	*Highway*	*Pipeline*	*Total*
1928	80 (approx.)	20 (approx.)			100
1938	51	46.1	2.9		100
1945	71.7	25	3.3		100
1953	56.9	28.6	8.5	6	100
1964	42.4	27	9	21.6	100

SOURCE: *Canada, One Century, 1867–1967.*

Plate 16.2 Until the last decade, most people saw the Rogers Pass during a once-in-a-lifetime vacation by rail. The building of the Trans-Canada Highway has opened the beauty of the Pass to many more, but the majority of travellers nowadays fly high above the Selkirks, almost unaware of the grandeur beneath them.

Boisclerc built a road between Quebec and Montreal. A new policy for roads, involving the notion of the "King's Road", did not appear until the nineteenth century. Not till 1946 could the Todd Medal be handed over in Victoria (B.C.) to the driver of a car which had come from Halifax (N.S.) without the help of the train and without a detour into American territory.

Canada did not, however, lack colourful episodes in the development of its road network. These include the Craig Road south of Quebec (1810); the import, in lieu of roads, of twenty-one camels for transport to Barkerville (B.C.) in 1862; the 485 miles of the Cariboo Trail (B.C.), built by the Royal Engineers between 1861 and 1864; wagon trains on the Prairies and, much later, tractor trains across the frozen lakes in winter in the North. During World War II, a track was built alongside the Canol pipeline from the Mackenzie to the Yukon and the long Alcan Highway between

Alaska and Canada was constructed in a single year (1942). In 1951, a road reached Chibougamau mining district in Quebec and in 1966, a track was built to facilitate the installation of a telephone link from Great Slave Lake to Inuvik on the Mackenzie Delta. In 1972, it was announced that work would begin on an all-weather highway running the whole length of the Mackenzie Valley. Before its completion, the Dempster Highway linking Dawson City to the Mackenzie Delta will be open.

It was after World War II, simultaneously with the rapid increase in cars and a rise in provincial highway expenditure, that the various road networks were linked to one another. Between 1949 and 1966, the Trans-Canada Highway came into being at a cost of almost one billion dollars, which was shared among all the governments concerned. The undertaking was the result of a national political decision: in economic terms the road is less transcontinental than the railway. In terms of traffic, it is more a series of individual sections linked up with one another than a through route from the Atlantic to the Pacific. The main traffic is in short sectors, such as from Vancouver to Hope, Banff to Calgary, Montreal to Quebec, and Corner Brook to St. John's.

Canada in one year spends almost $1.5 billion on the construction and repair of highways and city streets. The drivers of 8 million vehicles have available almost half a million miles of roads (800,000 kilometres), ten times more than the railways.

These improved highways, industrial maturity and decentralization, technical improvements in truck design and the speed and efficiency of road transport have all been factors aiding the development of trucking. Besides those areas where it dominates, such as within the towns and on the farms, it has made great strides in intercity transport. Because of regional economic disparities, distance and the existence of specialized methods of handling certain commodities (pipelines for petroleum, railways for cereals), the greatest amount of trucking is between the adjacent and highly populated provinces of Quebec and Ontario. Although competitive systems, rail and truck have some integrated services, like the piggyback (truck trailers on flatcars).

As highways use land and are a component of the landscape, they have to meet some environmental requirements. Therefore, the future route along the Mackenzie Valley should cope with terrain, vegetation, water, wildlife, waste material and archaeological concerns; engineering problems can no longer be considered alone.

Airlines

Although for bulk transport of freight the aircraft is insignificant when compared to other modes, it is a very different matter when the comparison is made in terms of total income from aviation, the transport of items such as mail and the contribution of aviation to transport in the Canadian North. So far as intercity transport is concerned, in 1964 aircraft provided more passenger miles than buses (36 per cent against 33 per cent) or railways (31 per cent). The aircraft has therefore become one of the major methods of transport. In 1971, air transport accounted for more passenger miles than rail and bus combined.

The experimental stage began early, with the flight of the Silver Dart in Nova Scotia in February 1909, and ended after World War I. In 1918, a mail service left Montreal for Toronto. Between the wars, the first flights into the North took place and airmail service between cities was established. The first aircraft in the Yukon arrived from Alaska on 16 August 1920 and the first in the Northwest Territories at Hay River on 22 March the following year. Around August 1925, a plane landed on the water of Hayes Fiord, Ellesmere Island. In 1927, aerial patrols were associated with federal surveys of Hudson Strait and helped to create the port of Churchill. In the remote North, aircraft are assuming the role once played by the railways on the Prairies. Indeed, it was aircraft which provided the first regular service of passengers, mail and freight to the mining town of Rouyn in the Abitibi.

In 1937, Canada established the basis of a national network with the formation of Trans-Canada Airlines (Air Canada since 1964). The war provided an opportunity for major development, and in 1943, the Canadian government, with the help of TCA, opened a transatlantic air route which, four years later, became a commercial operation.[5] In 1949, Canadian Pacific opened a service to Australia and New Zealand patterned on its sea routes. During the decade 1950–1960, domestic aviation profited both from the massive demand, which only air transport could meet, created by the building of northern radar lines and from great economic developments such as the iron ore of Quebec-Labrador and aluminum at Kitimat (B.C.). Dorval (Montreal) is the principal intercontinental airport in Canada but Malton (Toronto) handles the largest number of flights. Some cities, such as Calgary and Edmonton, Toronto and Montreal, are linked by an air bus service. Certain firms maintain executive aircraft. Fares are low, and for $120, in less time than it takes to cross the Atlantic, one can go from Montreal to Edmonton. In 1966, Air Canada alone transported 5 million passengers and extended its network to Moscow. In 1967, it carried an additional 1.3 million passengers, and in 1970, 7 million. Air Canada has become one of the world's largest airlines (the big U.S. companies excepted).

In view of the prospective growth in the number of passengers from Canada to Europe (already of the order of a million a year), Air Canada has taken options on several powerful jets. Around 1974, a new Montreal airport for intercontinental passengers will be in service. Toronto is planning one as well.

5. R. Garry, "Le développement de la ligne aérienne Trans-Canada", *Cahiers de Géographie de Québec*, No. 6, 1959, pp. 367–392.

Transport by Pipeline

As a component of the transport scene, pipelines represent a highly specialized mode when compared with other forms of transport handling all types of freight (see fig. 16.3). Their importance is due to the fact that petroleum and gas have become two of the chief driving forces in the national economy. Last century, some seepages of fuel oil were being tapped in Ontario, and in 1900, petroleum stood tenth in importance among mining products. During the construction of the Alcan Highway (1942), the petroleum reserves of the Mackenzie justified development; the Canol project involved an oil pipeline around 500 miles long, taking crude oil to a refinery at Whitehorse. The main developments, however, took place on the Prairies, with the development of the Turner Valley reserves in 1936, Leduc in 1947, Redwater in 1948, Pembina in 1953 and Rainbow in 1964. Crude oil reserves were estimated at nearly 10,000 billion barrels. These developments in the oil industry had to find both Canadian and American markets, and a series of pipelines was constructed, including the Transmountain, which reached the lower mainland of British Columbia in 1953. The main pipeline, the Interprovincial, built between 1950 and 1957 to link Edmonton and Toronto, had in 1970 a daily maximum capacity of 1 million barrels. It cost $0.48 per barrel to transport the black gold from Alberta to Sarnia, slightly less, despite the distance, than that via the 236-mile (379-kilometre) pipeline between Portland, Maine, and Montreal.

The youthful natural gas industry is more integrated than is the oil industry in the sense that the product is transported mainly by the extracting companies. These businesses are also more Canadian with regard to the supply and location of the gas pipelines. Even though discoveries go back to 1870, large-scale development only began after 1950. There are two main gas pipelines. The Westcoast pipeline, constructed between 1955 and 1957 and oriented north-south, runs from the Peace River country on the Alberta–B.C. border to Vancouver and the United States. The most important pipeline, the TransCanada, was completed in 1958 and runs from Alberta around the north of Lake Superior to Montreal and on to Sorel. The gas is used both for industrial and domestic purposes and for export.

Even though, between the St. Lawrence and the Pacific, these fuel pipelines do not form highly complex networks like the transcontinental railways, they are nevertheless just as essential a component of cross-Canada transport as are grain shipments, across an inland area where the alternative of water transport is not available. It is probable that gas and oil pipelines will be built across the Canadian

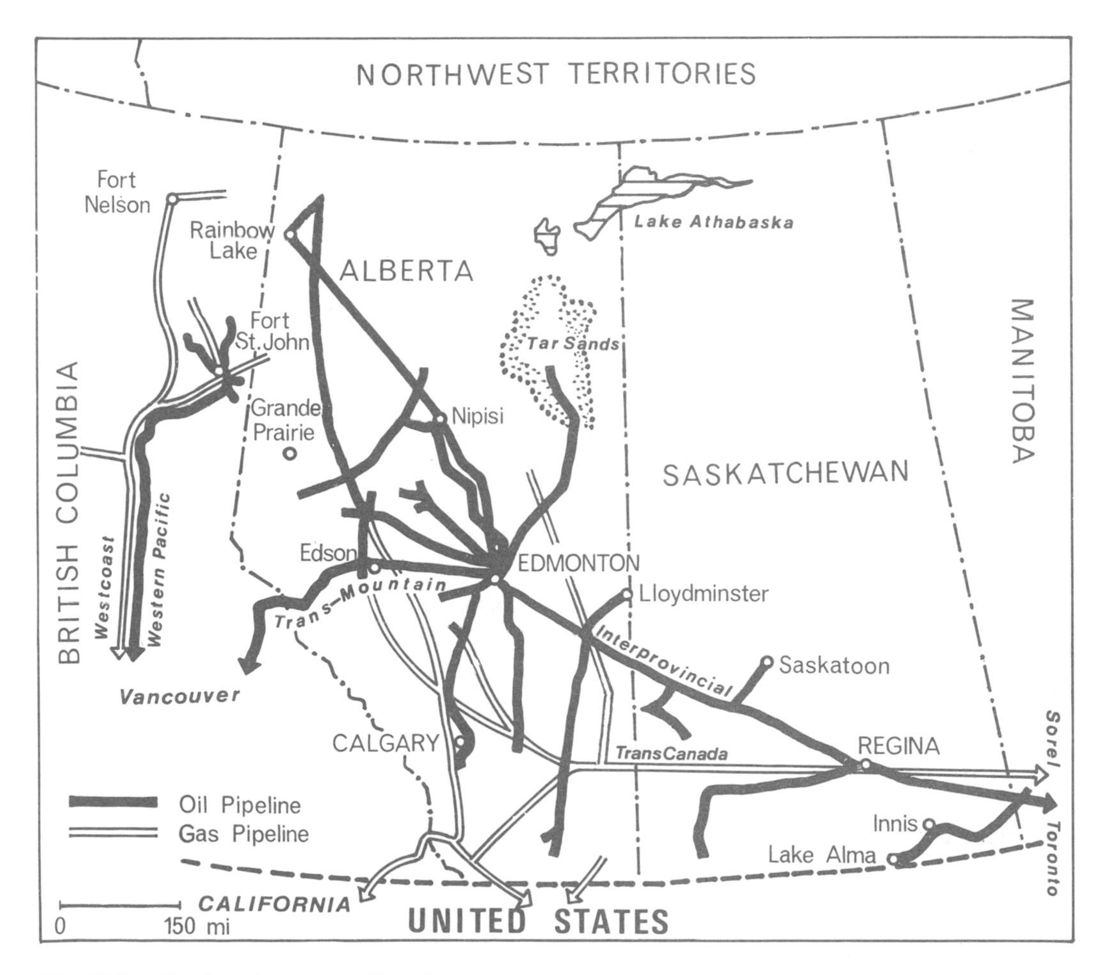

Fig. 16.3 Pipelines in western Canada.

Northwest to bring fuel resources from Alaska and arctic Canada to markets in southern Canada and the U.S.A. At least three problems have to be solved: Alaska competition, adequate protection of the environment, and Indian land rights.

Conclusion

Transport is so central to Canadian life that its economic aspects have been closely interwoven with internal political affairs. The Ontario canals, the Intercolonial and Canadian Pacific Railways, the Trans-Canada Highway, the disputes over pipeline routes, and winter navigation on the St. Lawrence are all reminders of the effect of commercial factors on Canadian identity. There is a continuous network of interactions which links the two main settled ecumenes, the Windsor-Lévis axis and the southern Prairies. Transport remains an absolutely essential function in Canadian society and economy.[6] The unity of the country, as well as its commerce, is dependent on transport and telecommunications.

Three chief needs remain to be met in the future: to improve traffic circulation in and around metropolitan areas; to provide better integration of and better competition among the chief components of transport; and to develop some form of transcontinental linkage across the Middle North.

External Trade

Trade is a major activity (see table 16.4). In 1964, the sums involved in external trade represented $813 for each Canadian, which gave Canada eighth place in the world, after Belgium, the Netherlands, Switzerland, Denmark, Sweden, Norway and Trinidad. In 1969, Canada took sixth position, with the volume of external trade representing $1,440 for every Canadian. The domestic market is comparable in size, the 1970 total being $28 billion or $1,304 per capita.

It is the strength of its external trade that is one of Canada's main attributes of power. In 1965, in absolute terms, Canada was one of the world's major trading nations. In 1969, Canada took sixth position, following the U.S.A., West Germany, Great Britain, France and Japan.

In relation to GNP, external trade is as important to Canada as it is to Great Britain (21 per cent in 1961–62), and more important than it is for West Germany (19 per cent), France (14 per cent) or especially the U.S.A. (5 per cent).[7] From this viewpoint, Canada is in a very different position from the U.S.A., where, in terms of spending power, the internal market is about fifteen times more important than it is in Canada. External trade is vital for the high standard of living of Canadians. As trade goes, so goes Canada.

Overall Development

Since about 1910, the Canadian share of international trade has developed at a more rapid rate than the average for the world. This is related to massive investments in Canada by several highly industrialized countries, as well as being the response of the Canadian economy to the demand for raw materials on the world market. Because of the high income derived from such trade, Canada has also been able to become a major importer.

World events have prevented this growth from being continuous. About a century ago, the opening of the British market to other nations made a reciprocity treaty with the

6. See [Canada], *Proceedings, Conference on Transport in the Arctic* (Ottawa: Department of Transport, 1971).

7. See M. G. Clark, *Canada and the World Trade*, study prepared for the Economic Council of Canada (Ottawa, 1964).

TABLE 16.4 Comparative Patterns of External Trade, 1963 (percentages)

	Manufactured Goods		*Industrial Materials (partly processed)*		*Primary Products*	
	Export	*Import*	*Export*	*Import*	*Export*	*Import*
Federal German Republic	78	31	11	16	10	45
Japan	76	21	16	7	8	72
United Kingdom	74	24	10	16	13	60
France	61	37	13	15	26	48
U.S.A.	58	33	8	20	31	45
Sweden	50	58	38	11	12	31
Canada	19	63	39	8	42	27

SOURCE: Based on M. G. Clark, *Canada and the World Trade.*

United States desirable (1854–1866). Confederation in 1867 created a customs union among British colonies in North America. During the last quarter of the nineteenth century, external trade suffered not only from a worldwide depression and from restrictive tariff policies, but also from the fact that it was for Canada a period of transition from the "wood, wind and water" era to the wheat era. The first three decades of the twentieth century revived external trade through cereal production, the adoption of preferential tariffs with Great Britain (1907), World War I and sales of paper and metals. The 1929 Depression led to a very strong contraction of business and forced a return to restrictive policies. After World War II, which was marked by a high level of exports, Canada adopted the GATT agreement encouraging freer and more efficient world trade.

Export and Import Products

From time to time (e.g., 1932–1950), the value of exports has exceeded that of imports. In the area of outgoing items, the furs, fish and wood products of the colonial period gave way to wheat and nonferrous metals at the end of the nineteenth century, and to paper about 1920. During the last decade, iron, oil and uranium together have accounted for more than 10 per cent of the total, which is equivalent either to wheat or newsprint. Compared to other industrial powers, however, Canada sells abroad relatively few highly manufactured products, though it imports a large amount. Although, like Sweden, Canada exports a high proportion of industrial materials, it is much more permissive in regard to the export of primary products than is Sweden. Compared to Great Britain or Japan, Canadian imports of those raw materials which are essential for industry are very small. It is true that exports of highly manufactured and fully finished goods have increased sharply, thanks to record sales of cars and automotive parts (see table 16.5). In 1966, manufactured goods had almost reached the value of agricultural products.

The relative growth of various items reflects the direction of the economy. Though Canada continues to depend on foreign imports for certain agricultural products, it has reduced such needs in the spheres of petroleum and

TABLE 16.5 Canadian Foreign Trade by Stage of Fabrication, 1950 and 1969 (by percentage of value)

Year	*Raw Materials*		*Fabricated Materials*		*End Products*	
	Export	*Import*	*Export*	*Import*	*Export*	*Import*
1950	30	31	56	30	14	39
1969	23	12	37	22	40	66

SOURCE: *Canada Year Book, 1971.*

TABLE 16.6 Major Components of Canadian External Trade, 1870–1965 (percentage)

	1870	*1910*	*1930*	*1950*	*1965*
Exports:					
(1) Agricultural products, including wheat	46	50	39	28	20
(2) Timber	33	14	5	9	6
(3) Newsprint			15	16	10
(4) Nonferrous metals	2	12	13	14	16
(5) Iron			1	2	5
Total (including other items)	100	100	100	100	100
Imports:					
(1) Textiles	24	17	15	12	7
(2) Agricultural and animal products	41	24	25	18	11
(3) Coal		8	6	6	2
(4) Crude oil		1	7	10	5
(5) Machinery and transportation equipment		4	14	26	35
Total (including other items)	100	100	100	100	100

SOURCE: *Canada 1867–1967.*

steel. As a result of developments in manufacturing, trade in textiles has lost much of its former importance, while machinery, transportation equipment and tools now form 54 per cent of the total value of imports. Compared to last century, Canada spends more on machinery and on acquiring the means of production, a desirable development in terms of a mature economy and an increase in productivity.[8]

8. See [Canada], *Canada 1867–1967.*

TABLE 16.7 Canada's Trading Partners, 1870–1969
(percentage by value)

	1870		*1900*		*1910*		*1930*		*1950*		*1965*		*1969*	
	Export	*Import*	*Export*	*Import*	*Export*	*Import*	*Export*	*Import*	*Export*	*Import*	*Export*	*Import*	*Export*	*Import*
U.K. and Commonwealth	41	58	59	25	56	30	36	21	21	21	20	11	12	10
U.S.A.	51	32	33	59	37	59	45	62	65	67	57	70	70	72
Others	8	10	8	16	7	11	20	17	14	13	24	18	18	18
Total	100	100	100	100	100	100	100	100	100	100	100	100	100	100

SOURCE: *Canada 1867–1967*, Annual Report of the Economic Council of Canada, 1964, and *Canada Year Book*, *1971*.

As table 16.6 shows, the relative importance in external trade of primary products and semi-finished goods has declined since 1950. In contrast, finished products (cars, aircraft, clothing, heavy machinery and equipment) have shown relative increases. This reflects the high standard of technology and a changing life style. The present situation is, however, very far from the ideal one of importing primary products and exporting finished goods, or even of not exporting primary products nor importing finished goods.

Direction of Trade

For two centuries, Canada's major trade partners have been English-speaking, but the most important country has changed. During the twentieth century, Great Britain's role has continuously declined, especially in regard to imports (see table 16.7). This has benefited the United States, which has become the chief customer and especially the chief supplier of Canada. Proximity, economic complementarity, and a variety of ties have proved stronger than the historical and sentimental links with Britain. In 1970, exports to the U.S.A. were almost eight times those to Britain (U.S.A., $10.6 billion; Britain, $1.4 billion). In external trade the latter (not including other Commonwealth countries) represents only about half the proportion contributed by other countries.

These other countries, however, represent too small a proportion of Canadian trade, when compared to the volume of trade with the U.S.A. Latin America, which in 1969 accounted for only 4 per cent of all Canadian imports and exports, could increase its trade links with Canada. Similarly, the massive sales of wheat to the U.S.S.R. and China could perhaps pave the way to more sustained trade relations with the socialist countries. Trade relations between Canada and France run at a very low level—only 1 per cent of the total value of external trade. The main exports to France were synthetic rubber, copper and wheat, while imports were mainly vehicles, wine and books. The recent growth of cultural, industrial and financial links between Quebec and France should lead to a growth of Franco-Canadian trade. Finally, the Pacific basin could, in the long run, be the "Ocean of the Future" for Canadian products. Between 1960 and 1970, exports to Pacific countries represented 7 to 9 per cent of Canada's total exports by value.

To increase its external trade, Canada must achieve two objectives. At the national level, it must keep its production competitive; that is, it must keep a close watch on costs and increase its productivity. At the international level, together with the other main trading nations and with developing nations, Canada must search for an overall multilateral balance which will optimize the flow of goods, of services and of payments. For the best Canadian relations with the European Economic Community, "the market must be cultivated assiduously."[9]

For several decades, the affluent society in Canada has been closely associated with the favoured sites in the nation, the urban areas. These are much more than industrial centres, labour reservoirs and consuming centres. Their metropolitan dominance is a controlling factor in national development.

---

9. See [Canada], *Canada Year Book, 1969*, p. 985.

Part Five

Urban Affairs

Although the urban population in Canada has been greater than that in rural areas for about the last fifty years, studies of urban geography have only recently begun to appear. Indeed, apart from Raoul Blanchard's pioneering work on Quebec in 1934[1] and the first topical urban geography in English written by Griffith Taylor of Toronto in 1946,[2] very few studies appeared before the middle of the century. All the urban regions have not yet been studied and there is still no comprehensive survey of the overall pattern, apart from demographic aspects.[3] Analysis of the urban fact is therefore well behind the growth of the fact itself (in 1972, only 6 per cent of the population was agricultural). Demographers have forecast that within fifteen years, half the Canadian population will be living in towns of more than 100,000 inhabitants. By the end of the century, more than one Canadian in three will live in Greater Toronto, Montreal or Vancouver.

The following urban survey consists of four parts, all devoted to the localized growth of population.

1. R. Blanchard, "Québec, esquisse de géographie urbaine", in *L'Est du Canada français* 2 (1935): 157–307.
2. See G. Taylor, *Urban Geography* (London: Methuen, 1946).
3. J. W. Simmons, "Urban Geography in Canada", *The Canadian Geographer* 11 (1967): 341–356.

Chapter 17

Population Concentration

Any study of urban population growth must take account of problems related to statistical definition. In 1951, in 1956 and again in 1961, the definition of urban areas was changed to bring the figures closer to reality. Reliance on legal definitions (i.e., the boundaries of incorporated municipalities) was abandoned and definitions became more realistic in population terms (1,000 inhabitants, minimum density) and in terms of location (fringes of areas with over 10,000 people). Such adjustments to 1951 figures would have increased the urban population by almost 700,000 people. Similarly, the definition used in 1961 excluded about 200,000 urban dwellers who would have been included in the broader definition used in 1956.

The Overall Pattern of Urbanization

These changes in the definition of the rural-urban boundary cannot obscure the overall pattern. In sixty years Canada has become a totally different place. Using the 1956 definition, 63 per cent of the population lived in rural areas in 1901. By contrast, 71 per cent of the 1961 population lived in towns; by 1966, the proportion was 73.6 per cent. The pattern of urbanization during the last hundred years can be divided into three general phases.

The first period, up to the time of the 1911 census, already had a well-developed urban life. From 1881 to 1911, the decennial percentage increase in urban population was in fact higher than that during the period 1911–1951, though this is probably due mainly to the low level of urbanization at the beginning of the period. In 1871, only 12 per cent of the total population lived in centres with more than 5,000 inhabitants. This early urban growth before 1911 is, however, an indication that urbanization did not depend on massive industrialization and that it took place during the settlement of the Prairies. A network of small service centres thus predates the larger towns based on manufacturing.

During the second phase, which lasted from 1911 to 1951, the pace of urbanization was slower, especially during the depression. After 1920, however, half the total population (which was around 9 million people) was nonrural, despite the fact that at the time, the number of manufacturing workers was only about 500,000.

The final period, characterized by a very high rate of urbanization, began during World War II and became a permanent feature after 1951. On the edge of Montreal, the growth of the South Shore exemplifies these tendencies to urban concentration.* During the decade 1951–1961, the total population of Canada increased by 4,200,000 but the urban population went up by 4,100,000; in other words, the growth of population has become almost entirely a matter of urban growth. In 1961, more than seven out of every ten Canadians lived in towns; in 1980 the proportion could be between eight and nine.

The pace of urbanization varies from one province to another. Base levels at the beginning of the century differed considerably. Saskatchewan and Alberta were then only 10 per cent urban, whereas in adjacent British Columbia, then the most highly urbanized province, the proportion was as high as 46 per cent. From 1961 census data, the provinces can be arranged in three groups:

(1) Those in which the urban percentage was higher than the national average of 71 per cent; that is, Ontario (79 per cent), British Columbia and Quebec. Since these are the three most important provinces in Canada, there appears to be a direct link between general economic development and urbanization.

(2) Six provinces, including three of the economically less prosperous Atlantic group and the developing Prairie Provinces, have urban proportions varying between 43 per cent and 63 per cent.

(3) In the agricultural province of Prince Edward Island and in the N.W.T., less than 40 per cent of the population is urbanized. Advanced urbanization is hence a localized phenomenon, although in the Territories there is at present a strong trend towards concentration of the population.

Similar spatial variations are to be found within individual provinces. In Quebec, for example, Metropolitan Montreal is 99.5 per cent urban, but the Gaspé–South Shore region is only 32 per cent urbanized and Abitibi-Timiscaming only 49 per cent. Even the decennial rate of urban growth is very variable in different parts of Quebec, varying from only 24 per cent in the Eastern Townships to 450 per cent on the North Shore and in New Quebec. This latter figure emphasizes that modern exploitation of the North is tending to bring about population concentration.

* The first wave affected Longueuil and St. Lambert between 1941 and 1947; a second, which lasted nine years, made Jacques-Cartier and Laflèche into working-class suburbs, dormitory towns for Montrealers or for less well off rural dwellers. The last wave, since 1956, has increased the population of this suburban region while bringing in a higher social class, for instance at St. Bruno and Préville.

The Urbanized Areas

What changes have taken place in the number of urban units in the different sized groups? Although the list of incorporated municipalities does not include all towns and villages, it nevertheless reflects the distribution of population among the main urban types. In 1871 the largest number of urban units contained between 5,000 and 30,000 inhabitants, while the least significant category was that of towns over 100,000 inhabitants. During the last thirty years of the nineteenth century, the establishment of numerous small centres made towns of 1,000 to 5,000 the most common; Canada was thus still a country of small towns. The shift in emphasis towards large centres only occurred in the twentieth century. By 1921, for the first time, it was the category of 100,000 and above which included the largest proportion of the urban population, 41 per cent. Since the end of World War II, the group containing 1,000 to 5,000 people has become the least significant. In short, 63 per cent of the urban population now lives in agglomerations of more than 30,000 people (see table 17.1), in contrast to

TABLE 17.1 Number of Urban Centres (more than 15,000 people) in Different Sized Groups by Province and Major Region, 1961

	Population (in thousands)								
	15–25	*25–50*	*50–100*	*100–200*	*200–300*	*300–400*	*400–500*	*500–1,000*	*Over 1,000*
Newfoundland		1	1						
P.E.I.	1								
Nova Scotia	1	2	1						
New Brunswick	1	1	1						
Total for Atlantic Provinces	3	4	3						
Quebec	17	17	5	1					1
Ontario	18	12	7	2	2			1	
Manitoba	1	4			1				
Saskatchewan	1	1	1	1					
Alberta	2	2			2				
Total for Prairies	4	7	1	1	3				
British Columbia	1	1	1			1			
Territories									
Totals	43	41	17	4	5	1		1	1

SOURCE: Based on *Canada Year Book, 1965.*

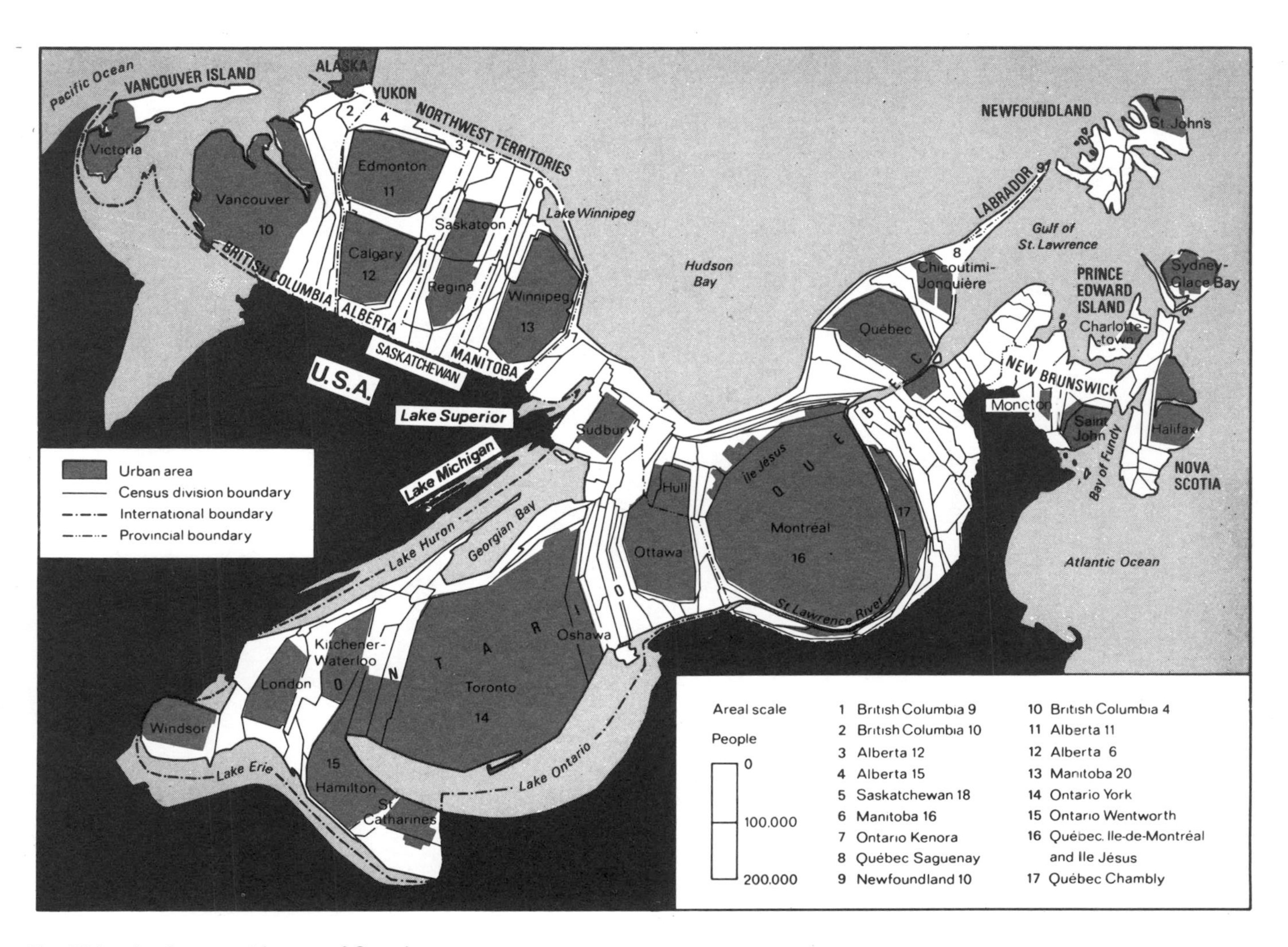

Fig. 17.1 Isodemographic map of Canada.

only 34 per cent in 1871. Over the same period, urbanization has meant the spatial extension of these centres.

This analysis of urban population in terms of the size of the town distinguishes the usual regional variations. As might be expected, the Atlantic Provinces and the Prairies, when compared to the other major regions of Canada, have a greater proportion of their urban populations in centres containing less than 10,000 people, whereas Ontario is well represented in the groups with over 30,000 people. The "urban continuum" of Ontario is much better developed than that of Quebec, which contains many small centres, few of moderate size, and the supercity of Montreal. In 1961, Quebec had no town with a population between 172,000 and 1,191,000 inhabitants. The Atlantic Provinces had no towns over 100,000, while the Prairies had none over 300,000. Quebec, Ontario and even British Columbia show a wide scatter of sizes. Fifty thousand inhabitants seem to form a significant barrier in several provinces. Urban structure is thus far from uniform across the provinces of Canada (see fig. 17.1).

Metropolitan Growth

Has the recent growth of population benefited the metropolitan centres more than the smaller units? Y. Kasahara, comparing the rates of growth of metropolitan regions with those of other urban regions during the decade 1951–1961, showed that rates were higher in the larger regions in every province, except for British Columbia.[1] It was in Manitoba and in Alberta that the two rates showed the greatest contrast, 93 per cent compared to 15 per cent. The seventeen metropolitan areas accounted for 60 per cent of the 1951–1961 decennial increase in the Canadian population. One Canadian out of two now lives in a census metropolitan area (core and fringe). One in four Canadians lives in one of three such zones. The ecumene is becoming ever more concentrated.

The Canadian metropolitan areas are not, however, maintaining the same rates of growth relative to each other (table 17.2). Because of their present size, Montreal, Toronto and Vancouver will remain of first importance, but Toronto's more rapid growth suggests that it will displace Montreal as the largest unit. Revived by the oil industry, the two major cities of Alberta, Calgary and Edmonton, have shown the greatest rates of increase, 98 per cent and 94 per cent. Winnipeg will be displaced from the fourth position by these cities and by two others that are growing rapidly, Ottawa and Hamilton. For industrial reasons Windsor, with a decennial increase of 18 per cent, and Quebec (30 per cent) seemed to be losing their momentum, although Greater Quebec has been recovering again in recent years.

This metropolitan growth can also be assessed in relation to the outward spread of the population, a characteristic feature of the North American city. Schnore and Petersen have shown that from 1871 to 1921, the rate of population increase was much more rapid in the main urban core than in the suburban areas (table 17.3).[2] Since then, the opposite has always been the case, although there now seems to be something of a decline in the rate of outward movement. This urban spread represents on the one hand an areal separation between home and place of work and on the other a reflection of changes in the means of transport. From 1911 to 1961, the population distribution between town centres and suburbs has differed considerably from one urban area to another. Despite the availability of land within

1. Y. Kasahara, "A Profile of Canada's Metropolitan Centres", *Queen's Quarterly* 10 (1963): 303–313.

2. L. F. Schnore and G. B. Petersen, "Urban and Metropolitan Development in the United States and Canada", *The Annals of the American Academy of Political and Social Science* 316 (1958): 60–68.

TABLE 17.2 Comparative Rates of Growth in the Largest Urban Areas, 1951–1961

Urban Areas in Order of Size of Population in 1961	*Rank in Terms of Absolute Increase in Population, 1951–1961*	*Rank in Terms of Percentage Increase in Population, 1951–1961*
Montreal	1	4
Toronto	2	3
Vancouver	3	7
Winnipeg	8	10
Ottawa	6	5
Hamilton	7	6
Quebec	9	11
Edmonton	4	2
Calgary	5	1
Windsor	12	12
Halifax	11	9
London	10	8

SOURCE: 1961 Census.

TABLE 17.3 Comparative Growth (%) of Cities and their Suburbs

Decade	*Centre*	*Periphery*
1871–1881	39	13
1901–1911	83	21
1921–1931	29	33
1941–1951	16	46
1951–1961	16	97

SOURCE: 1961 Census, and based on Schnore and Petersen (1958).

their cities, Edmonton and Calgary even more (478 per cent) have had the greatest rate of peripheral growth. Even the City of Toronto has lost some of its population with consequent reduction of population density, while the suburbs have grown by 114 per cent, a rate higher than the Canadian average of 97 per cent. Some towns have had practically no suburbs, including Regina and Saskatoon.

Metropolitan growth is therefore more recent than urbanization, and it began when half the Canadian population was already urban. The growth of the suburbs is absorbing an increasing proportion of the population of metropolitan areas; the rates have grown from 25 per cent in 1941 to 45 per cent twenty years later. In 1961, of the seventeen metropolitan areas, seven were in the Province of Ontario.

The Urban Nature of the Population

The traditional division between rural and urban is no longer adequate; in 1966, the respective proportions (26 per cent and 74 per

Plate 17.1 Although southern Ontario is an inconvenient distance from first-class skiing, Nathan Phillips Square in front of Toronto's city hall provides opportunities for careful experiments in another winter sport.

TABLE 17.4 Population: Urban, Suburban and Rural (1961 and 1966)

Population Groups	*Inhabitants*		*Percentage of the Canadian Population*	
	1961	*1966*	*1961*	*1966*
Farm	2,072,785	1,913,714	11.0	10.0
Rural nonfarm	3,465,072	3,374,407	19.0	13.0
Urban:				
Metropolitan fringe	3,698,296	4,530,360	20.3	24.0
Metropolitan nonfringe	4,465,690	5,104,383	24.7	26.6
Urban nonmetropolitan	4,536,404	5,092,016	25.0	26.4
Total urban	12,700,390	14,726,759	70.0	77.0
Total	18,238,247	20,014,759	100%	100%

SOURCE: 1961 and 1966 Census.

cent) were unsatisfactory as indicators of the real division of the population. There are in fact five main classes, as shown in table 17.4.

This analysis of the 1966 statistics identifies two important characteristics: the rural, non-agricultural group represents a community whose functions are not those of agriculture and whose outlook and habits orientate them towards the urban dweller. In some respects, therefore, 90 per cent of the Canadian population is urban or urban-oriented. Secondly, the table shows that two-thirds of the Canadian population live outside the metropolitan cores. In terms of size of urban units, therefore, Canada is meso-urban rather than hyperurban.

According to estimates shown in table 17.5 for Canada in 2001, the country would have eight great cities with over one million in population size, none of these in the Atlantic Provinces. Ontario will still dominate the urban polarization with 42 per cent of the total population of the twelve agglomerations studied.

TABLE 17.5 Forecasts of Urban Population for 12 Major Centres, Year 2001

Toronto	6,510,000
Montreal	6,374,000
Vancouver	2,482,000
Ottawa	1,616,000
Winnipeg	1,614,000
Edmonton	1,223,000
Hamilton	1,201,000
Quebec	1,178,000
Calgary	937,000
London	674,000
Windsor	577,000
Regina	438,000

SOURCE: N. H. Lithwick, *Urban Canada. Problems and Prospects.*

Toronto, although first, will stay in competition for primacy with Montreal. Vancouver will reach a size comparable to the large Canadian cities today. Ottawa should be in fourth position, following the big three.[3]

Here, only population size is considered. However, this single component is a reliable indicator of economic activities.

The agricultural community has experienced a rapid decline; in 1931 it still comprised 31 per cent of the total population. The depression led to a pause in the abandonment of agriculture and from 1931 to 1941, Quebec and Alberta even showed an increase in the population living on farms. By contrast, the war speeded up rural depopulation and even the wheat province of Saskatchewan lost 116,000 people from agriculture. The decade 1951–1961, particularly the second half, took more than 500,000 people off the farm, half of them in Ontario and Quebec; all the provinces except Newfoundland were severely affected. It seems inevitable that the national proportion in agriculture will fall below 6.5 per cent. In the various provinces, the present proportion varies from 33 per cent in Prince Edward Island and 25 per cent in Saskatchewan to 2 per cent in Newfoundland. Alberta and Manitoba, with 15 per cent, remain above the Canadian average, while Quebec (6 per cent) and Ontario (7 per cent) are already at that level.

Population changes in the future may show a slight contraction of the agricultural sector, a modest increase in downtown living, a slowing down in the pace of metropolitan suburban development, and, lastly, a greater increase in nonmetropolitan urban regions than in the rural nonfarm sector.

3. See N. H. Lithwick, *Urban Canada. Problems and Prospects.*

Chapter 18

Space and Urban Landscapes

This chapter is concerned with the site, situation, factors influencing development, external appearance and internal character of the urbanized areas, viewed in historical terms.

The urban area is a concept which is not clearly defined. Before 1956 it meant the population of a municipal unit designated a town or an incorporated village. Since 1961 this legal emphasis has been replaced by a demographic and functional definition: the urban area includes those places which have a population group of at least 1,000 inhabitants or which are situated within large metropolitan areas otherwise defined. These definitions are not ideal because urban units of 1,000 inhabitants and particularly the areas within metropolitan influence include enclaves which are much less urban, whether judged in terms of population or of function.

The Establishment and Development of Urban Centres

Forms of Development

A useful distinction can be made between coastal and riverine towns, those on railway and highway routes, commercial centres, manufacturing towns and political and administrative capitals.

Urban functions are seldom defined once and for all and, with the passage of time, towns reflect the effect of man's initiative or hesitation.

Urban studies are hence characteristically dynamic. Take, for example, the Ontario Lakeshore. In his study of population growth, J. Spelt identified four types of centre.[1] The first type includes the mainly industrial towns in which there has seldom been a pause in development; Toronto is the prime example but Oshawa, Peterborough and Barrie are others. The second type consists mainly of service centres such as Lindsay and Collingwood, which got off to a good start but which have not been able to maintain the same pace of growth (some of these areas have even declined from time to time). A third type of evolution, characterized by prolonged stagnation, is characteristic of some small towns. The last group comprises those small towns which, after marking time in the second half of the nineteenth century, have expanded as a result of industrial development.

In some cases, functional development seems to have followed a cyclical pattern, with periods of prosperity following periods of depression. Joliette (Quebec), for example, has gone through two complete cycles between 1850 and 1950. Further, even if we confine ourselves to the main factors affecting urban growth or stagnation, there is an enormous tangle which must be unravelled; rarely is a single factor solely responsible. For instance, many factors interacted in the growth of Sault Ste. Marie. They include: its site on a large navigable waterway used by the fur trade; rapids which both interrupted river traffic and now provide hydro power; American and Canadian iron ore upstream and coal downstream; the nearness of the Canadian Shield, which restricted agricultural development but encouraged the lumber industry; the international boundary; a market in Ontario for steel and paper; and lastly, the distance from other urban centres. Key factors are always found in different combinations. The strategic site of Quebec City is no longer important and the citadel, which as it happens was never used as such, has now only an historic and tourist importance. The city's position, once considered inland in relation to the Atlantic, was for a long time outside the main heartland of the continent. Once the capital of Canada, Quebec has become a provincial town, so much have economic factors favoured her rival, Montreal. Container developments in the port have, however, recently revived the commercial economy of Quebec City.

New sets of events have even led to shifts in the established order. Frequently a new factor may lead to the development of a second urban nucleus, and ultimately, the creation of a conurbation. It is precisely because of this dynamic equilibrium of factors that it is difficult to accept models of an urban cycle in which an urban unit passes through definite stages in its development. Some towns nowadays are indeed fully developed at birth, like Thompson in northern Manitoba.

Form and Pattern of the Built Environment

The urban area represents only an infinitesimal portion of the area of Canada; even in Ontario, it is only 0.6 per cent. Only Greater Montreal and Greater Toronto cover extensive areas. At the local scale, however, spatial problems are of major importance, and in the course of the next decade, half a million acres must be converted to urban uses.

In Ontario and Quebec

In southern Ontario the urban pattern consists of two main groups:

(1) There is a major concentration including Toronto-Hamilton and Kitchener, with a total population in 1971 of about 3,350,000.

1. See J. Spelt, *The Urban Development in South-Central Ontario* (Essen: Van Gorcum, 1955).

About 100 miles (160 kilometres) long, and enclosing the western end of Lake Ontario, this urbanized area has not yet reached Oshawa in the east nor Brantford in the south. The planners forecast that this "Mississauga" or "Toronto-centred region" will extend for 160 miles (257 kilometres) in thirty years' time.
(2) More modest are the three separate centres of Ottawa (602,510, including Hull, Quebec), Windsor on the international border (258,643) and London in the interior (286,011).

These two groups of cities and their immediate surroundings contain two-thirds of the urban population of Ontario, although there are a number of centres with 10,000 to 50,000 inhabitants around Greater Toronto but outside the conurbation.

In southern Quebec, four major groups are recognized, which again take in two-thirds of the urban population of the province.

(1) The Montreal area (2,743,208 inhabitants in 1971) is marked by a fragmentation, which is the result both of uncontrolled spread of settlement and a plethora of municipal units, shorelines and the main physical routes. Though the main urban core is on Montreal Island, compact urban areas have developed on Ile-Jésus as well as on the north and south shores of this fluvial archipelago. Its overall size requires emphasis: the 1981 forecast is for about 3.5 million inhabitants.
(2) Though Greater Quebec has, like Montreal, a site which is divided by the river and which extends out along the main highways, the urban agglomeration is much more coherent, and in particular there is much less of an abrupt break between the centre and the satellite areas. The regularity of this urban landscape reflects a slower rate of growth than that of Montreal. From 1951 to 1961, Quebec had one of the lowest rates of population increase of Canadian centres comparable in size.
(3) The conurbation of Middle Saguenay (Chicoutimi-Jonquière), with 133,703 inhabitants, forms the third urban zone of Quebec. It consists of several competing centres around Chicoutimi, which is only marginally the main unit.
(4) To these must be added the isolated towns of the Eastern Townships and Mauricie.

Three Canadian Metropolises: Montreal, Toronto and Vancouver

Montreal

For more than a century, Montreal has been the leading urban centre in Canada in several respects. Its pride in being the second city of the French-speaking world does not cause it to forget that it is also the only major French-English city in the world.[2]

Its checkerboard pattern stretches along the St. Lawrence and now surrounds the forested refuge of Mount Royal. In the townscape, the first impression is that of a core of skyscrapers, seen to good advantage from the Look-Out on the Mountain. This core is bounded by the port and by Amherst, Sherbrooke and Guy Streets; that is, it is contained between the Mountain, the Lachine Canal and that part of the St. Lawrence River between the older bridges of Victoria and Jacques Cartier. In this sector, land is expensive. This core structure differs from that of Toronto, the heart of which seems much less dominant relative to surrounding areas. Further, the orientation of Toronto is more at right angles to a waterfront which is less developed. The central business district (CBD) in Montreal serves many functions; in addition to the commercial and financial nodes

2. P. Y. Denis, "Montreal: bilan décennal d'une morphologie en transition", *Revue de Géographie de Montréal* 25 (1971): 281–300.

which are now tending to coalesce, 22 per cent of the floor space in 1962 was residential, 12 per cent industrial and 7 per cent was owned by institutions. In recent years this varied urban environment has been spectacularly developed around Place Ville Marie, Place Canada, Place Bonaventure and Place Victoria.

In Montreal, the varying densities of population generally reflect the period of settlement and the presence of industry. The area north of Mount Royal (locally known as the east) is much more densely populated than any other part of the metropolis: it constitutes the sector around Lafontaine Park (280,000 people), with extensions along Maisonneuve, Rosemont and Villeray Streets. It is a fairly homogeneous residential sector, characterized by terraced, three-storey houses. Equally densely populated, and equally French-speaking, are the areas of St. Henri and Verdun southeast of the Mountain. Around and between these foci are zones with lower densities. They include: a predominantly English-speaking semicircle around the south and west of Mount Royal; Rivière-des-Prairies in the northwest; Longue-Pointe to the north; and the South Shore to the east. In these areas, single- and two-family homes predominate, with a generally brighter appearance than those in the centre.[3]

The style of building emphasizes both the period of settlement and the degree of cosmopolitanism. A few old "French" houses, doomed either to systematic demolition or to renovation, in very good taste, are surrounded by a variety of "English" styles and more especially by monotonous two- or three-storey brick buildings. Since World War II, low, American-style houses (bungalows and split-levels) have become as popular in the suburbs as have the high-rises in the centre. From a cultural point of view, Montreal is more a town of contacts between several ethnic groups than a melting pot, with four large universities, the French-language Université de Montréal and Université du Québec, and the English-language McGill University and Sir George Williams University.

Raoul Blanchard has shown clearly that the port and the Lachine canal have been responsible for the location of significant industrial areas in the southeast of Montreal;[4] similarly, the railways too have led to the development of manufacturing on either side of Mount Royal. Manufacturing developed rapidly in wartime in eastern Montreal between the shipping berths and the oil refineries. For a long time the river seems to have been an obstacle to the establishment of industry on the South Shore, and from 1940 to 1960 (i.e., until the opening of the St. Lawrence Seaway in 1959), the latter consisted only of dormitory suburbs. The postwar period saw the development of diversified industrial activity throughout the Montreal metropolitan region. Industry and service functions are, however, poorly related to the residential population and this makes it necessary for much of the labour force to travel long distances to work, which in turn has made transport improvements necessary.

As a result of its national importance, Montreal has an impressive transport system. Though the rail network encircles Montreal, the freight yards are mainly located in the zone from St. Lambert through Verdun and Lachine to Côte St. Luc. These large areas devoted to transport are extended westwards by the Dorval International Airport, which is more than two miles in width. The port function extends from the Lachine canal to Montreal-East and the St. Lawrence Seaway runs along the South Shore. At this transition point between ocean and inland waterways, the large terminal and transit port handles ships of all kinds. The urban transport system was extended by the building of the first two lines of the Metro in time for Expo '67.

3. L. Beauregard, "Population nocturne et diurne à Montréal", *Revue de Géographie de Montréal* 28 (1964): 290–296.

4. See R. Blanchard, *Montréal et sa région* (Montreal: Beauchemin, 1953).

The site of the international and world exposition on the islands in the St. Lawrence and the new roads which have been built will probably create closer links between the citizens and the attractive waterfront; up till now, apart from the visitors to Ile-Ste.-Hélène, the St. Lawrence at Montreal has been more like the Thames in London or the Plate at Buenos Aires than the Seine in Paris.

Toronto

Halfway between Windsor and Ottawa, Toronto has a central location in southern Ontario (fig. 18.1), the most powerful region in Canada, and is really the political, commercial, industrial, financial and cultural capital of the province. The name *Toronto*, which means 'meeting place', is fully justified, as is its nickname, "queen city". Administratively, Metropolitan Toronto (2,086,017 inhabitants in 1971) contains five boroughs and a city. In terms of population, the City of Toronto (712,786 inhabitants in 1971) can be distinguished from the semicircle of the five boroughs of Etobicoke, North York, York, East York and Scarborough (1,373,231 inhabitants). Since 1945, Greater Toronto has grown enormously, and a more massive shape has replaced the rather finger-like appearance it had at the end of the war. Toronto is more than a provincial capital; it has played a major role in the economic development of the Prairies and of the Canadian Shield. It is the centre of English Canada, and in some respects, the first city of all Canada.

The Toronto land-use map is very complex. The main industrial areas are along the railways and beside the harbour, which has been considerably extended into the lake. Hence the land used by industry and as docks separates the lake from the rest of the city. The main industrial nodes are this lower town, the port, Leaside, Weston, Swansea, New Toronto, Mimico, Parkdale, Dupont and East Toronto. Since 1950, a relocation policy has encouraged the development of industry in suburbs such as Rexdale.

In the lower part of the city, the main services are concentrated close to the historic Yonge and Bay Streets. Toronto is far and away the main financial centre in Canada. The CBD, which consists of an amorphous mass of buildings of different heights, is now extending outwards at the expense of older residential areas. This area has a J shape; it has clearly defined limits and consists of three principal groups: retail stores in the north, offices in the south, and municipal buildings in the west. For the last ten years, half the new business floor space in metropolitan Toronto has been located outside this traditional centre. This has led in turn to a similar dispersal of service functions so that the heart of Toronto now accounts for only 20 per cent of retail sales; the effectiveness of the competition from suburban shopping centres is obvious.

Just north of the CBD, and around it, are the residential areas. As in many Canadian cities, such as Regina, the railway, which was once on the edge of the built-up area, later restricted urban growth. In Toronto, these barriers are increased by the presence of the beach of a former glacial lake, and especially by deep ravines. In 1955, the land-use map of Toronto still showed a large gap in the settled ecumene along the Don Valley. Since the war, the expansion of suburban homes has been the characteristic feature of Greater Toronto's growth. Compared to Montreal, the population density of Toronto is much less; it is also more uniform and less focused on the central business district. Toronto lacks a Mount Royal to exert a restrictive influence and to raise densities in the central area. At any rate, the westward sprawl of housing has preceded that in Montreal. Toronto, until very recently "very British", in the words of Kerr and Spelt,[5] is becoming cosmopolitan and more alive.

---

5. See D. Kerr and J. Spelt, *The Changing Face of Toronto*.

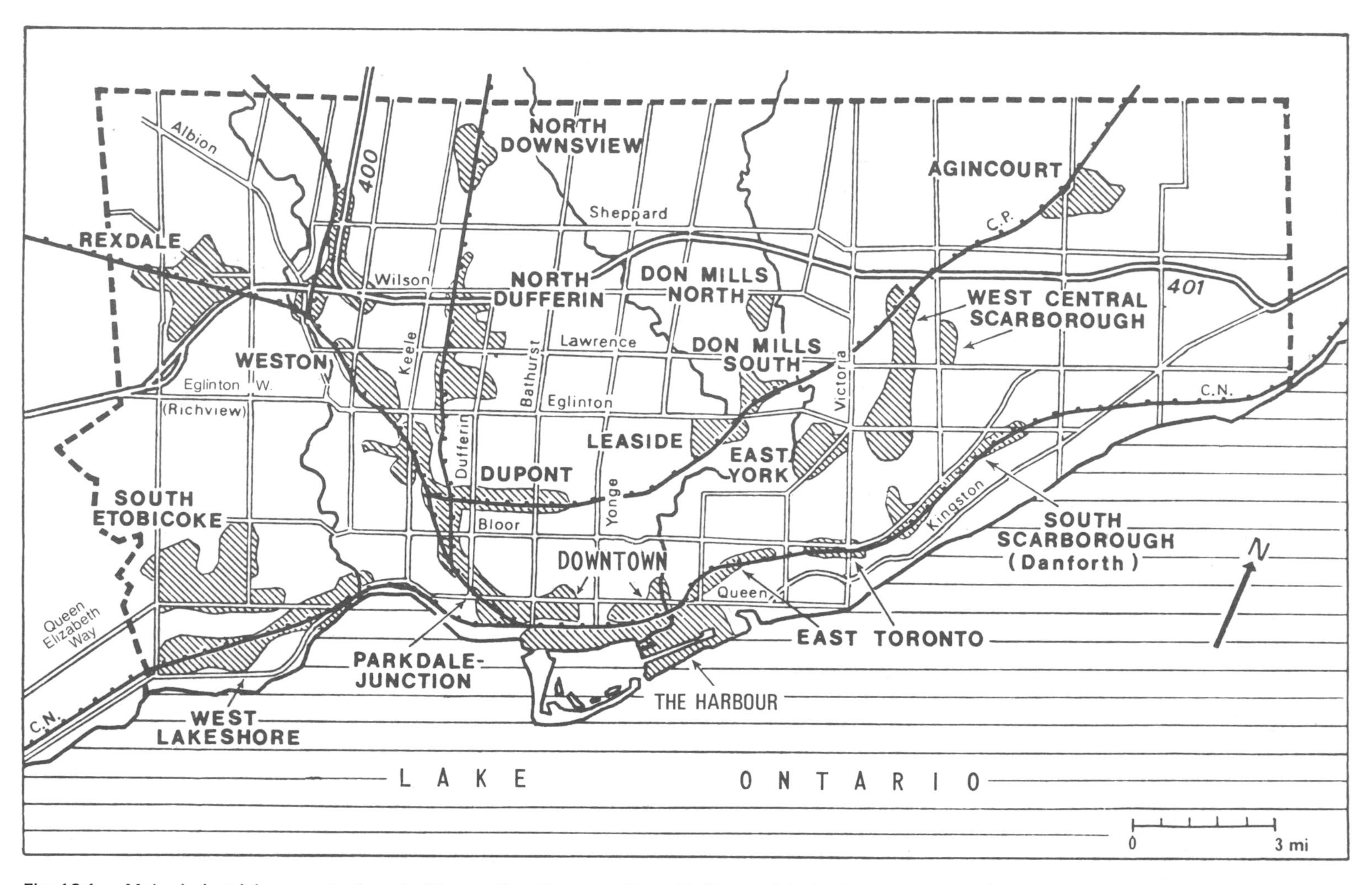

Fig. 18.1 Major industrial concentrations in Metropolitan Toronto. (From D. Kerr and J. Spelt, *The Changing Face of Toronto.*)

Despite the large parks, such as that which houses the Canadian National Exhibition, city planners are creating more green space for the ordinary city dwellers, something which is much less common in French Canada. The 1947 report was an important landmark in active town planning. Ahead of Montreal, the capital of Ontario provided itself with an underground transit system and with major highways such as the Queen Elizabeth Way and the Macdonald-Cartier Freeway (Hwy. 401).

In southern Ontario, there is a Toronto-centred region 50 to 60 miles wide and covering an area between Peterborough, Kitchener and southeast Georgian Bay; inside this urban watershed, Toronto influences the evolution of all the communities.

Greater Vancouver

Greater Vancouver (1,082,352 inhabitants in 1971), a single focus in its province, as Greater Winnipeg (540,262) is in Manitoba, is the most important urban centre in western Canada (fig. 18.2). It has become the first millionaire city in Canada outside the St. Lawrence Basin. With a magnificent, if fragmented, site, the town owes its expansion as much to the presence of a low plain sited on the edge of the continent as to its coastal location. It vies for the position of being the largest port in Canada.

The site of the main economic activity is determined by the ridge between Burrard Inlet and False Creek. Port facilities are mainly located along the channels of Burrard Inlet, which is more suitable than the various arms of

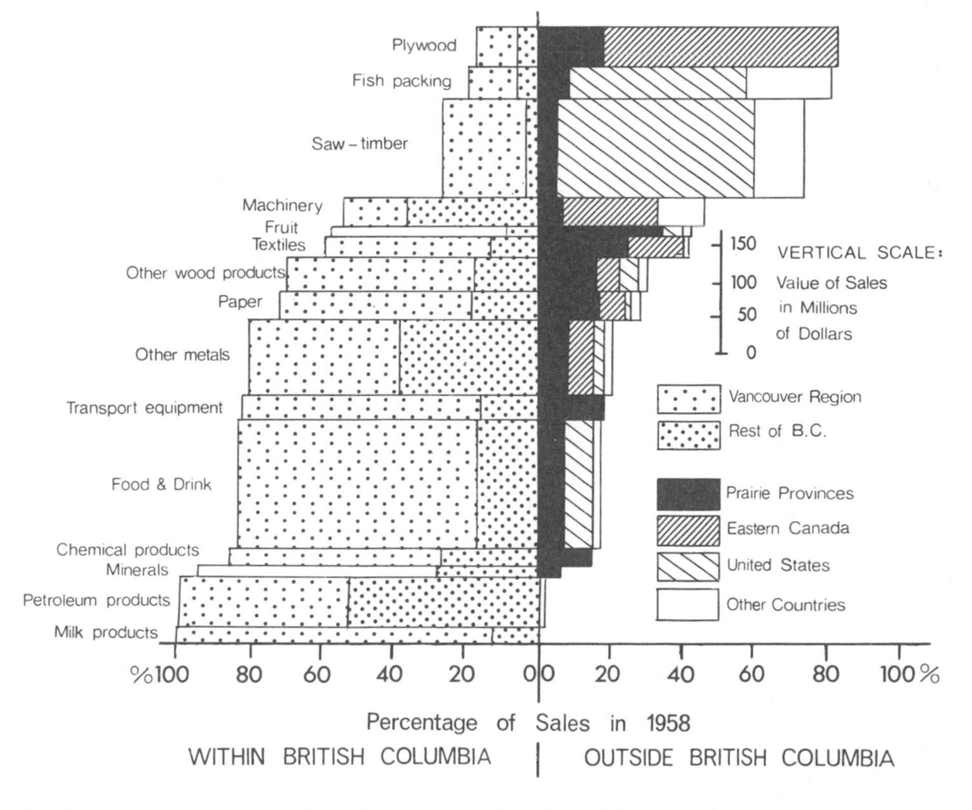

Fig. 18.2 **Markets for Vancouver's industrial output. (After P. D. McGovern.)**

the Fraser River because of the absence of silting. Immediately adjacent, around the embayment of False Creek, are the main commercial and industrial areas and the termini of transcontinental railways, both Canadian and American. Because of the availability of water and open land, other industrial sites have developed farther south, along the north arm of the Fraser and its islands. Some manufacturing industries have also been established at New Westminster and on the north shore of Burrard Inlet.[6] Commercial activity is solidly anchored in the Georgia peninsula, accessible by long bridges. Elsewhere, New Westminster and North Vancouver are second-order centres.

At the beginning of the century, large areas devoted to single-family homes spread out over the peninsula of Vancouver and then to North Vancouver. Thirty years later, growth southwards had reached the north arm of the Fraser and the old river centre and rival of New Westminster, which is joined to the centre of Vancouver by a direct road, Kingsway. There are sharp contrasts between the older gridiron patterns, the modern suburbs with their curving streets in the Capilano Valley, and the almost rural environment of the Fraser Delta. Near the commercial centre is the most important Chinese quarter in North America after that in San Francisco. The residential population is evenly distributed, apart from a slight concentration around the downtown peninsula; the area's schools are similarly almost uniformly distributed.[7] Away from the satellite centres of the West End, Fairview and New Westminster, the normal house type is single family. The greater part of Vancouver has a building density of only 10 to 25 per cent, which leaves large areas for lovely gardens, the pride of the city.

6. P. D. McGovern, "Industrial Development in the Vancouver Area", in R. L. Gentilcore, ed., *Canada's Changing Geography*, pp. 182–198.
7. See [Canada], *Urban Analysis Series. Vancouver, British Columbia*, report of the Geographical Branch (Ottawa, 1965).

More than most other urban areas, Vancouver, like Victoria, has some large and magnificent parks, especially Stanley Park beside Lion's Gate Bridge and, in the west of the city, the campus of the University of British Columbia, which includes a Japanese garden. Greater Vancouver includes forested areas such as Burnaby Mountain Centennial Park (the site of the new Simon Fraser University) and open spaces on Sea Island (airport) and Lulu Island. Other recreation facilities include numerous marinas, mountain chairlifts such as Hollyburn and the road along the Capilano Valley. "The functions of Vancouver in Western Canada are more that of a Los Angeles or San Francisco than that of the nearby American cities of comparable size, Seattle and Portland."[8]

The Regular Pattern of Urban Areas

The uniform appearance of the built environment is a striking feature. At first sight, the repetitive patterns seem to be due to carefully controlled planning. This is not in fact the case: the ideal form of an urban unit is not to be equated with a street pattern in rectangles or regular curves. In most cases, the geometric layout is a result of land survey. In the St. Lawrence Lowlands of Quebec, several municipalities are arranged in parallel sequence back from the river. In the Eastern Townships and in the Abitibi counties, some of the central market towns have boundaries which parallel their parish. The towns of Basse-Beauce display characteristics both of the natural setting and of the old seigneurial divisions in their layout. Winnipeg, Portage-la-Prairie and Brandon are based on township layouts. In British Columbia, near the mouth of the Fraser, urban growth at Cloverdale and Haney closely follows the pattern of land survey. This geometric form of city

8. W. Hardwick, "The Individual Province", *The Geographical Magazine* XLIV (1972): 611.

Plate 18.1 The narrow peninsula of downtown Vancouver, seen from the south. The Granville and Burrard Street bridges link the downtown with the residential area south of False Creek. Older single-family homes were overshadowed during the 1960s by high-rise apartment buildings facing Sunset Beach and English Bay. Behind this West End lies the wooded area of Stanley Park and behind the park, on the north side of Burrard Inlet, the rapidly growing suburbs of West Vancouver and North Vancouver extend up the hills and along the Capilano Valley.

plan is reinforced by the gridiron pattern of streets, the first prototypes dating from 1749 in Halifax and perhaps even from the seventeenth century in Montreal by Dollier de Casson. This gridiron is found all over Canada in towns of all ages: Charlottetown, Quebec outside its walls, Joliette, Montreal, Toronto, Oshawa and the majority of Prairie towns like Regina, which followed the homestead subdivisions. Most of Vancouver is neatly divided into rectangles of equal size.

There are few exceptions to this ubiquitous system of squares and rectangles, though mention should be made of the haphazard pattern of settlement in Newfoundland. Some other exceptions include a few Ontario towns with a diagonal pattern of streets. The centres of both Guelph and Goderich, built around 1830, consist of avenues radiating from a central focus. Unfortunately, a rectangular pattern of streets soon encircled this radiating form. Another radial form arises simply through the triangular convergence of groups of townships which have different orientations, as, for instance, the star-shaped plan of Stratford and, less clearly, that of central Winnipeg. Parts of towns like Nanaimo, B.C., have a fan-shaped plan; a similar form in the newer parts of Sarnia, Ontario, provides a contrast to the gridiron of the old town.

The dominance of such regularity, despite valleys, hills, winds, the need to commute, neighbourhoods and even the attitude of the inhabitants, has frequently been criticized since the first conference on urban development at Winnipeg in 1917. Despite this, much recent growth has followed the same pattern.

House Types

There are four main types of home—separate houses, apartments, duplexes and row housing; the first of these remains the most popular.[9] The long tradition of single-family dwellings is such that in Vancouver in 1961, 75 per cent of households were in separate houses. Types and ownership of urban homes vary from one province to another. In Quebec, there is much rented accommodation and much more apartment living than in Ontario. The recent trend towards apartment buildings in English Canada and an increase in suburban separate houses in French Canada are tending to diminish this contrast in the built environment. "Having represented one out of every four units started in the early 1960s, apartment-type dwellings have more recently been accounting for one out of every two and for three out of every four in Toronto and Vancouver."[10]

Canadian towns are also differentiated by age; the period of settlement accounts for the newness and more open character of the towns on the Prairies and even more those in the Middle North, in contrast to the old, congested town centres of the St. Lawrence Lowlands. House types, open space, and population density vary considerably from place to place.

Patterns of Urban Development

The metropolitan areas consist of a series of differentiated and interrelated spaces. From the centre to the edge, one can usually distinguish in turn a core (or the original node), a suburban area (which is unevenly developed but which generally includes some satellite centres), and lastly the outer ring (where the influence of the principal town gradually diminishes). The simplicity of this concentric scheme hides the complexity of the links between the centre and the inner and outer hinterlands, because the centre is not the only hub, nor are the suburbs the only source of population. The intraurban and the periurban regions interact and even compete with each other. For almost a hundred years, before and after 1900, railways and roads on the one hand, brick, stone and cement on the other, encouraged the development of the town centres. During the last twenty-five years, in contrast, the great increase in both population and parking, together with the aging of downtown areas, has encouraged a trend towards decentralization: the periurban ring has been the area in which recent urban growth has been concentrated. From 1951 to 1961, the suburbs of Montreal absorbed 282,000 or 77 per cent of the 364,000 additional inhabitants in the whole metropolitan area. The revival of the centre is therefore a problem, given the strength of this centrifugal movement.

Central Area Development

Some general observations can be made, even if they are based on specific examples. The centre takes some time to react. Economic revival

9. A. Bailly, "L'essor de l'habitat collectif dans les villes canadiennes", *The Canadian Geographer* XV (1971): 127–140.
10. *Canadian Housing: Achievements and Challenges*, The Bank of Nova Scotia Monthly Review (Toronto, 1972), p. 1.

during the war years was able to conceal the reality of peripheral expansion from the businessmen who were firmly established downtown. Some downtown areas lost certain of their traditional functions; for example, Université Laval moved outside Quebec City, and the old Rue St. Jean in Quebec City experienced a decline after 1930. Moreover, in eastern Canada some intraurban industries were deliberately wound up or were not rebuilt after fires.

The most forward-looking centres have reacted by attacking two of their main problems: traffic and housing. The car has put enormous pressure on town centres. In 1959, 378,000 vehicles left or entered downtown Toronto between 6:30 and 11:30 A.M.; at the junction of Hwys. 401 and 400 in Toronto, the daily summer traffic reaches 120,000 vehicles. Ring roads as in Winnipeg, linear routes such as Laurier in Quebec, or elevated highways such as Granville in Vancouver have been built. In Toronto, the Gardiner Expressway in the south, Highway 27 in the west and Highway 401 to the north enclose the metropolitan area, while an internal network based on the Don Valley Parkway and Spadina Avenue surrounds the business district. Even in the prairie cities, which already have wide streets, multilevel garages have replaced decayed property. In downtown Montreal, from 1949 to 1962, of all forms of land use it was the area devoted to parking which experienced the greatest growth, partly due to the demolition of old properties. The new parking lots are not merely essential for business but also for homes. During winter nights, cars may not be left on city streets lest they hinder snow clearance.

The spatial problem is particularly serious in town centres developed before the car. The larger stores in the centre have been able to keep a minimal clientele only at the price of providing parking space, especially since the suburban shopping centres provide vast amounts of free parking. The numerous cities located on rivers or waterfronts often lack adequate bridges; for example, in Montreal, Quebec and Vancouver.

In the last few years, the development of the road network in Greater Montreal, to serve more than one million vehicles, has been astonishing. It was made possible by a bold municipal policy and by Expo '67. After the Laurentian autoroute in 1958 came the opening of the Metropolitan Boulevard around 1960. Subsequently came the construction of an enormous interchange linking the Trans-Canada Highway, routes towards Ontario, and the Eastern Townships autoroute via the new Champlain Bridge. In the north of the city, the remarkable Lafontaine Bridge and Tunnel in 1967 extended both the Trans-Canada Highway and a new expressway towards the northwest. In the same year, after the Concordia Bridge, came the opening of the partly excavated Decarie Boulevard. Towards the east, the Metropolitan Boulevard was extended in the direction of the new autoroute towards the northeast. In total, Montreal Island has a network of 100 miles (160 kilometres) of expressways, with numerous access roads.

Elsewhere in Canada there have been similar achievements. In St. Vital, near Winnipeg, major elevated traffic interchanges have been provided at the intersection of the principal highways. The same is true along Route 499 across the lower Fraser Valley, south of Vancouver, not to mention along the impressive Highway 401 around Toronto.

The pedestrian has not been spoiled to anywhere near this extent: for him only a few malls, like that in Ottawa, and also two underground systems have been created. In 1966 Toronto added to its north-south subway line an east-west line along Bloor Street from Woodbine to Keele. The Montreal Metro, begun in 1962, opened at the end of 1966. In the heart of Montreal the railway and metro stations, office buildings and the main hotels are linked by air-conditioned underground walkways, extending for 2 miles (3.2 kilometres) and packed with shops. The pedestrian can satisfy all his needs without going outside, an undeniable benefit during winter blizzards. Edmonton is among some other Canadian cities building similar facilities.

The obsolescence of downtown housing areas has frequently been studied. Taking Montreal as a whole, 34 per cent of the buildings are over forty years old, according to the city planning department; in the Petite-Bourgogne (east of Atwater), 72 per cent of the buildings were more than sixty years old. A reasonable life for such buildings, given the hard climate, is around fifty-five years. Urban housing policy includes the elimination of slums, the preservation of historic properties, the restoration of decayed but still essential buildings and, finally, comprehensive development. In the centre of Toronto, according to a 1944 report, only 16 per cent of the residential space was considered "sound", 32 per cent was "vulnerable", 50 per cent was "declining" and 2 per cent was "very bad". In the face of such conditions, downtown areas have had to renew their urban fabric. Inevitably to provide office space, tall blocks (more than forty storeys in the large cities), covered with facings of aluminum or plate glass, have pushed vigorously up through the lower slabs of stone and cement of an earlier generation. In Toronto, for example, alongside the Royal York Hotel, are two tower blocks of fifty-six and forty-six storeys; Vancouver is building similar structures for the federal and provincial governments. Though the new buildings have been built as offices and banks, old houses have been converted into shops. Commerce has thus displaced residences in the downtown areas of many cities.

Here and there, as a result of land scarcity, there have sprung up enormous "machines for living", subdivided into a large number of small apartments. Although among these properties there are some which are the result of essential urban renewal (the Dozois schemes in Montreal and Regent Park in Toronto), the majority consist of high-rent apartments for single people and small families. The town centres, which for twenty years have seen their residential functions decline as service functions have increased, have in this way been able to renew the tradition of housing some of their workers. In some cities, whole neighbourhoods have been renewed, including Ottawa, which previously followed a rigid policy of demolition. In Vancouver, beside the small but magnificent English Bay, elegant apartment properties are springing up among the older houses. In the centre of some cities, the revival of movement has been so spectacular that the area has been described, rather unsuitably, as a "boom town". But before this new phenomenon of "return to the town", there was an outward spread.

The Periurban Environment

For the last twenty years, one of the dominant geographical facts in Canada has been the peripheral growth of already existing towns. These recently urbanized suburbs are a real laboratory.

The forces converting rural areas into suburbs have been irresistible; although in the last twenty-five years they have been altered, they have remained intense. The serious economic depression left a great demand for jobs. In 1937, Canada was a small country. Wartime industry helped the cities, which hence attracted the workers and deprived the rural areas of adequate labour. Since urban renewal had come to a stop during the depression, the town centres could not house all their immigrant workers. The suburb that could be reached by private car offered land at a lower and lower cost, farther and farther from the centre. This dispersion met the general demand for single-family housing, as well as a latent nostalgia for a countryside left behind so recently and a desire to get away from "life on an apartment balcony". The political system, not ready to accept planning, did not restrict outward growth or the private enterprise system and did not discourage profitable speculation on building sites. The lack of organization and the belated growth of metropolitan governments could therefore not control the centrifugal flight of population. Peripheral growth took place as fast as designs could be drawn, and the plans were concerned more with building lots than with urban growth as a whole.

Suburban growth led to conflicts between construction and older activities. Since the majority of towns were originally service centres for nearby rural populations, urban growth clearly affected agricultural land. The degree to which the latter resisted the assault has received little general study.

A. D. Crerar posed the following problem: how many acres of land have been taken out of agricultural production for every 1,000 additional citizens in towns of more than 100,000 inhabitants?[11] The author first noted that his calculations could not be applied to Halifax, where there is little adjacent agricultural land. Further, the suburbs on lands of marginal productivity, such as those in Ottawa and Quebec City, have reacted differently from those where soils are better. Around the urban centres of London, Winnipeg, Toronto-Hamilton and Montreal there has been an average loss of 382 acres (154 hectares) of agricultural land for each 1,000 new inhabitants.

The gross theoretical density of population in the new peripheral urban areas would therefore be only 2.6 inhabitants per acre (8.5 per hectare). These values are low when compared with the norms proposed for a loose suburban area: 9.2 persons to the acre (23 to the hectare). There is also an enormous difference in density to that in town centres: in 1961 in Montreal, there was an average of 154 persons to the acre (383 to the hectare) in census tract 135 (near St. Denis). The low density of this new periurban settlement has led to its being described as absolute and relative waste of land. The waste of land is most serious where expansion takes place at the expense of land whose agricultural production is significant for all of Canada. This has been particularly important in the Niagara fruit belt, the output from which cannot easily be replaced elsewhere in Canada. The pattern of

11. A. D. Crerar, in *Resources for Tomorrow*, supp. vol. (Ottawa, 1962), pp. 209–225.

TABLE 18.1 Loss of Agricultural Land Compared to Metropolitan Population Growth

Metropolitan Region	*Population Growth 1951–1961*	*Loss of Agricultural Land (acres)*	*Acres Lost per 1,000 Growth in Population*
Ottawa	55,585	55,656	1,001
Quebec	41,250	41,121	1,000
London	26,671	12,231	458
Winnipeg	56,134	20,741	383
Toronto-Hamilton	409,152	156,259	382
Montreal	284,986	106,525	374
Windsor	23,943	4,605	192

SOURCE: Based on A. D. Crerar (1962).

present building development could cause the Niagara fruit zone to disappear as an agricultural region; good planning, on the other hand, could provide space for both peaches and people.[12] Similarly, the Montreal area is extending outwards at the expense of the best agricultural region in Quebec. In 1941, there were still 10,000 people on the farms of the Montreal archipelago and 35 per cent of the enterprises derived their principal income from vegetables and fruit. This rural character has virtually disappeared. Taking the Montreal Plain as a whole, experts forecast a deficit in meat, eggs, cheese and fruit. At the other end of the country, Greater Vancouver, to which labour commutes from a forty-mile radius, competes with the agricultural life of the Fraser Valley. Urbanization destroys much cultivated land, often without using the land intensively for urban needs (table 18.1).

Not all this land is built over. A pilot study of four towns in southern Ontario around London distinguished the true expansion or built-up area from the "urban shadow", in which the land only loses its agricultural uses.[13] The result showed that:

(1) The urban shadow affected an area much greater than the area actually used for building. For the year 1960, the respective areas per 1,000 inhabitants were 542 acres (212 hectares) and 106 acres (42.9 hectares) for Kingston, and 190 acres (76.8 hectares) and 97 acres (39.2 hectares) for London.

(2) Two-thirds of the area in the urban shadow consisted of good-quality soils, but the shadow discourages cultivation because it breaks up lots, leads to short-term leases which discourage long-term cultivation practices, and raises site values.

(3) The urban shadow does not necessarily develop in a manner proportional to population change. Lindsay (12,500 inhabitants in 1960), which had only grown by 2,000 people in ten years, had an urban shadow stronger than that of London, which grew from 98,000 to 175,000 during the same period—but London has a planning commission.

(4) The areas of the urban shadow affected by speculation fetched high land prices which ultimately hindered building expansion. Although land speculation—a "national sport" —stimulates business, it becomes a plague when carried to extremes. Discussing the Toronto region, a planner spoke of "the devil of land speculation".[14] Eighty per cent of the land taken out of agriculture on the edge of Montreal was being held for speculative gain. J. B. Racine has estimated that 30 per cent of farms in the rural municipalities around Montreal are rented.[15] The urban shadow has a variety of forms: regular areal spreading, isolated spots similar to terrain conditions at Kingston, ribbon development along the highways to Stratford. In Canada, real estate speculation is nothing new: Regina suffered from it even before the city was established. The control of speculation would save suburban dwellers astronomical sums.

A great many spatial problems are caused by this excessive and insufficiently planned growth. It certainly appears that there has been an extraordinary waste of periurban land; the same number of people could have been accommodated very satisfactorily in more limited areas.

---

12. R. R. Krueger, "The Disappearing Niagara Fruit Belt", in R. R. Krueger et al., *Regional and Resources Planning in Canada* (Toronto: Holt, Rinehart and Winston, 1963).
13. J. H. Smith, in *Resources for Tomorrow*, supp. vol. (Ottawa, 1962), pp. 179–209.
14. L. Gertler, "The Price of Living in Developing Ontario", *The Geographical Magazine* XLIV (1972): 464.
15. J. B. Racine, "Exurbanisation et métamorphisme peri-urbain", *Revue de Géographie de Montréal* 21 (1967): 313–341.

Through rapid development, low-density suburbs have been created, raising the cost of services per family. Experts agree that the outer urban ring has not been used in the best way. This wastage of land increases the distances which must be travelled and the loss of time is accompanied by a greater wear on biological capital. Paradoxically, while consuming much land, the outer fringe has not provided abundant reserves of water, nor agricultural lots, nor public recreation areas and green spaces, nor lines for rapid transit. It appears that before long, we will have to replan, at great expense, urban land which could have been better used in the first place.

The general appearance of these sprawling peripheral neighbourhoods contains little of aesthetic value. Apart from some so-called model subdivisions, such as the Parc Falaise in Quebec City or Don Mills in Toronto, the outer fringes of the major cities lack originality; they are not particularly adapted to local conditions and they are frequently very monotonous. "See one suburb, see them all." Granted, there are many boxes of tulips and strips of lawn, but the grass does not replace the tree that was cut down to reduce the cost of site preparation. To see these bare subdivisions, with little atmosphere, low density, geometric street pattern and houses that are often alike, one is led to believe that the pseudo-urbanism which presided at their conception was oriented more towards construction than towards society. Some suburban extensions even consist of substandard housing, as on the South Shore of Montreal and especially the small suburb of Whitehorse.

The periurban advance of settlement has taken place in administrative disorder. There has been an overlap of government functions: for example, the federal government controls the airports, the provincial government looks after the intercity highways and municipal governments are responsible for zoning. Even more troublesome is the multiplicity of autonomous municipal authorities. Although the metropolitan zone forms an organic whole, the overall effect of urbanization cannot be appreciated by the local councils, and they are often ill-equipped to try. Cooperation between the parent city and the satellites is frequently weak or even nonexistent. The city of Montreal has often complained that it pays for services, such as fire protection, which profit all its neighbours. Some towns, it is true, have established joint administrative relationships with their suburbs; Greater Edmonton has in this way been able to avoid unnecessary expense in the management of roads and common services. It was not until 1965 that the municipalities of Ile-Jésus were amalgamated. A Greater Winnipeg unit came into existence early in 1972. In general, the spread of suburbs has taken place in conditions of extreme fragmentation of power; metropolitan commissions have been more institutions of consultation than of government. Given such administrative disorder, it would indeed have been miraculous if urban growth had been attractive, economic and functional.

The two areas, intraurban and periurban, must be related in an operational plan. The National Capital Commission works in this way in Colonel By's town of Ottawa. It focuses on six principal objectives: the increase of open space, the seat of government, the green belt, the railway network, Gatineau Park and the redevelopment of the central area. The great population increase of the last decade has created a population of over 600,000 inhabitants in the two-province region of Ottawa-Hull. It seems that Ottawa will attain a position more appropriate to its status as capital.[16]

---

16. See A. Coleman, *The Planning Challenge of the Ottawa Area*, Geographical Paper, No. 42 (Ottawa, 1969).

Chapter 19
Urban Functions

In the preceding chapter we considered two of the four classical functions of cities, habitation and circulation. There are various ways in which the other two functions, production and services, can be studied. Individual towns naturally develop a general reputation for certain activities: for example, Trail, B.C., is associated with metal working, and Esterhazy, Sask., with potassium; Chalk River, Ont., has an atomic reactor; Corner Brook, Nfld., is a paper town and Inuvik, N.W.T., is an administrative centre. Sometimes this image involves a distinctive but temporary activity: Calgary is the home of the Stampede; The Pas, Man., has a trappers' festival and Quebec City a winter carnival. Although such impressions are justified, they are only qualitative; we must look at the question in other ways.

Methods

Some urban functions can be assessed in terms of indices or measures of output; for example, Toronto was responsible in 1961 for two-thirds of all share transfers on the country's stock exchanges. Such measures, however, are not appropriate for all urban functions; they are only significant in the biggest centres and they only assess the urban function from a business viewpoint. The study of individual buildings, or even of the whole urban area, provides another perspective. For example, a radar site, a port installation, or a university campus like that at Waterloo are indicators of specific activities. The fact that Toronto has 68 per cent of the head offices of the major Canadian mining corporations demonstrates the dominant position of this city in the development of the Canadian Shield. In some defence settlements in the North which are too small in population and extent to qualify as urban, it is the nature of the physical structures which justifies the conclusion that their main significance is not rural in character.

A still more geographical form of analysis, perhaps, is provided by the land-use map. The function of a town is represented in the landscape by the pattern of buildings and the use of space, and by a spontaneous (or sometimes planned) arrangement of crossroads, streets and neighbourhoods. On such maps, for example, dormitory towns stand out very clearly. An integrated survey of buildings, sites and activities can facilitate the detailed analysis of the dominant function; for example, along MacLeod, a major shopping street in Calgary, it is possible to sort out from among the variety of activities those that are tied to a central location, those that serve the passing trade, those that serve the street itself and those that provide a special function.

One method of analysis takes as its basis the fact that employment is a measure of urban function. In other words, the occupational diversity of the population is a measure of the character of urban life. Early research of this kind was limited to calculating the percentage of the labour force engaged in each sector of the economy, but about 1956, the form of this analysis was improved. It can be argued that it is not the total labour force that determines the functional hierarchy of urban centres, but only those of it who serve the external market; those whose product is consumed locally are not used in the definition of urban function. To provide a threshold beyond which employment can be considered to serve more than local needs, minimum percentages have been established for each main industry. According to J. W. Maxwell, for Canadian towns with more than 100,000 inhabitants in 1951 the standard values were: manufacturing, 12.4 per cent; retail trade, 10.3 per cent; construction, 5.8 per cent; education and health, 5.7 per cent; personal service,

5.4 per cent; transport, 5.1 per cent; wholesale trade, 3.0 per cent; public service, 3.0 per cent; finance, 2.5 per cent; business, 1.2 per cent; public utilities, 0.8 per cent; recreation, 0.4 per cent.[1] The percentages are slightly different in smaller cities.

If, in an individual industry or city, the percentage of workers exceeds the threshold, the additional amount is declared extraurban. It is "excess employment" as compared to "basic employment" and it is a measure of the urban function. Such calculation defines the function in three ways:

(1) It identifies the dominant function of a town in terms of the occupation that has the greatest excess of employment, as, for example, manufacturing in Arvida, Quebec.
(2) It identifies the distinctive function of an urban unit as compared to other units. Of all the towns in Quebec, Arvida is the one with the highest excess employment in manufacturing; its counterpart in Ontario is Oshawa. For any specific centre, the dominant *absolute* function and the dominant *distinctive* function are not always the same. For example, in the provincial capital of Regina, government services have the largest excess employment, but it is the number of employees in wholesale trade which most clearly distinguishes the city from other urban areas.
(3) Finally, the character of the urban function is specified by an index of specialization that expresses the total deviation of each activity from the theoretical model. An isolated capital like Charlottetown, whose external activities are very similar to the basic model, is therefore considered to be unspecialized. The majority of centres have values of specialization below 5 (e.g., Montreal, 3.92 and Toronto, 3.51).

1. J. W. Maxwell, "The Functional Structure of Canadian Cities: A Classification of Cities", *Geographical Bulletin*, No. 7, 1965, pp. 79–104.

It is this technique, the determination of the real and distinctive importance of urban functions beyond local needs, that is used in the following classification of Canadian towns.

Specific Functions

J. W. Maxwell has made a detailed study of the functional index of the main urban centres in 1951.[2] Since cities exist in a dynamic equilibrium, even though the overall situation which he described is still true the subsequent growth of tertiary and quaternary employment has altered the significance of certain dominants.

Absolute Dominance

Of the eighty centres studied, sixty-one were primarily concerned with manufacturing, six were mining centres, five were transport centres, five primarily provided government services, two were primarily concerned with retail trade, and one was an education centre. Hence, as a function of employment in nonlocal needs, 76 per cent of these Canadian cities were manufacturing centres. In some respects, however, this fundamental character is not as widespread as it appears. The majority of urban areas in which the dominant function is manufacturing are mainly located along the axis from Windsor to Lévis. Of the forty-five towns in southern Ontario and in the Montreal Plain, forty-three are in this manufacturing category, whereas only one-third of the urban areas in other regions have manufacturing as the dominant function. Manufacturing is therefore far from being scattered all over the country.

The degree to which manufacturing is the dominant component is also important; the significance of a specific function is usually indicated by whether it accounts for more or less than 50 per cent of the excess employment. Of the sixty-one manufacturing towns, twenty-five,

2. *Ibid.*

or 41 per cent, are below this 50 per cent level, and they are mainly located outside the Windsor-Lévis axis. This characteristic is also significant in table 19.1, which shows thirty-three urban areas characteristic of different provincial conditions. Away from the main manufacturing axis of the country, other industrial towns of less significance include St. John's, Thunder Bay, Winnipeg, Saskatoon, Calgary, Edmonton, Vancouver and Trail. The same table shows that one-quarter of these thirty-three towns have a dominant function other than manufacturing. For Rouyn-Noranda it is mining, for Charlottetown retail trade, and for Brandon and Moncton it is rail transport. For the four capital cities of Regina, Victoria, Halifax and Ottawa, it is government services.

Distinctive Dominance

The order of precedence changes if distinctive functions are considered. The tables clearly show the difference in importance of the two dominant functions. Manufacturing accounts for three-quarters of the thirty-three examples in table 19.2 on an absolute scale but it accounts for no more than 33 per cent in relative terms. In this analysis, diversification appears to be considerable, and wholesale trade, financial services, transport and government services are all quantitatively significant.

Functional Classification of Urban Areas

In terms of the degree of specialization and the excess employment in manufacturing and in wholesale trade, J. W. Maxwell has grouped the eighty Canadian cities into five classes:

(1) thirty-one specialized manufacturing centres, mainly localized along the Windsor-Lévis axis: e.g., Welland and Trois-Rivières
(2) seventeen regional centres where trade is important, such as Edmonton, Regina, Fredericton and Rimouski
(3) twenty multipurpose towns, such as Medicine Hat, Thunder Bay and Kingston

TABLE 19.1 Comparison of Dominant and Distinctive Functions in 33 Representative Cities

<table>
<tr><th rowspan="2">Function</th><th rowspan="2">Dominant*</th><th colspan="2">Distinctive</th></tr>
<tr><th>Number of Towns</th><th>Percentage</th></tr>
<tr><td>Manufacturing 1 (more than 50% excess)</td><td>42</td><td rowspan="2">10</td><td rowspan="2">33</td></tr>
<tr><td>Manufacturing 2 (less than 50% excess)</td><td>33</td></tr>
<tr><td>Government</td><td>12</td><td>3</td><td>9</td></tr>
<tr><td>Transport</td><td>6</td><td>3</td><td>9</td></tr>
<tr><td>Retail trade</td><td>3</td><td>1</td><td>3</td></tr>
<tr><td>Mining</td><td>3</td><td>1</td><td>3</td></tr>
<tr><td>Finance</td><td></td><td>4</td><td>12</td></tr>
<tr><td>Wholesale trade</td><td></td><td>6</td><td>18</td></tr>
<tr><td>Education</td><td></td><td>1</td><td>3</td></tr>
<tr><td>Construction</td><td></td><td>3</td><td>3</td></tr>
<tr><td>Public utilities</td><td></td><td>1</td><td>3</td></tr>
</table>

* as a percentage of the 33 towns.

SOURCE: Based on J. W. Maxwell (1965).

TABLE 19.2 Dominant and Distinctive Functions of 33 Characteristic Cities

		Dominant Function						Distinctive Function (Ranked)												
Cities	*Provinces*	*Retail Trade*	*Mining*	*Manufacturing 1*	*Manufacturing 2*	*Government*	*Transport*	*Construction*	*Retail Trade*	*Wholesale Trade*	*Education*	*Mining*	*Finance*	*Manufacturing & Processing*	*Recreation*	*Business Services*	*Government*	*Personal Service*	*Transport*	*Public Utilities*
St. John's	Nfld.				+				4	1	5						2	6	3	
Sydney	N.S.			+					4	3				1	5			6	2	
Halifax	N.S.					+		3		5			6				1		4	2
Moncton	N.B.						+		2	3	5		4				6		1	
Charlottetown	P.E.I.	+							1	4	6		5		2			3		
Arvida	Que.			+							2			1						
Quebec	Que.				+			1		2	4		3				5			
Trois-Rivières	Que.			+				3			4			1						2
Shawinigan	Que.			+				2						3						1
Sherbrooke	Que.			+				2		3	4			1						
Sorel	Que.			+										1						
Montreal	Que.			+						2			1							
Rouyn	Que.		+					4		5		1			2	6		3		
Ottawa	Ont.					+							2		3		1			
Kingston	Ont.				+				3		1		4		5	6	2			
Oshawa	Ont.			+										1						
Toronto	Ont.			+						3			1			4				2
Hamilton	Ont.			+										1						

TABLE 19.2 (cont'd.) Dominant and Distinctive Functions of 33 Characteristic Cities

		Dominant Function						*Distinctive Function (Ranked)*												
Cities	*Provinces*	*Retail Trade*	*Mining*	*Manufacturing 1*	*Manufacturing 2*	*Government*	*Transport*	*Construction*	*Retail Trade*	*Wholesale Trade*	*Education*	*Mining*	*Finance*	*Manufacturing & Processing*	*Recreation*	*Business Services*	*Government*	*Personal Service*	*Transport*	*Public Utilities*
Kitchener-Waterloo	Ont.			+									1	2						
London	Ont.				+			3		4	5		1				6			2
Windsor	Ont.			+										1						
Sarnia	Ont.			+				1						3		2			5	4
Sault Ste. Marie	Ont.			+										1	3				2	
Thunder Bay	Ont.				+			3		4					5				1	2
Winnipeg	Man.				+					1			2		4				3	
Brandon	Man.						+			2	4		3		5				1	6
Regina	Sask.					+			6	1			2				3		5	4
Saskatoon	Sask.				+				5	1	3		2				6		4	
Calgary	Alta.				+			5		1			2		3		6			4
Edmonton	Alta.				+			1		2			4				5		3	
Trail	B.C.				+									1	2	3				
Vancouver	B.C.				+					1			2			4			3	
Victoria	B.C.					+					3		2		4		1			
Total		1	1	14	11	4	2													

SOURCE: Based on J. W. Maxwell (1965).

TABLE 19.3 City Groupings, 1961

Group	*Main Characteristics**	*Examples*
1	Eastern industrial frontier	Rouyn, Edmunston, Timmins
2	Industrial frontier (metals)	Arvida, Sault Ste. Marie, Trail, Sudbury
3	Southern Ontario	Barrie, Trenton, Stratford, Guelph, Peterborough, Welland
4	Older service centres in Ontario, Prairies, Maritimes	Regina, Medicine Hat, Prince Albert, Charlottetown, Moncton, Edmonton, Winnipeg, Vancouver, London, Kingston
5	Smaller Quebec centres	Grand'Mère, Sorel, Drummondville, Joliette, Saint John, Cornwall
6	Affluent suburbs of Montreal and Toronto	Leaside, Mount Royal, Westmount
7	Central cities	Montreal, Toronto
8 and 9	Heterogeneous characteristics	Halifax, St. John's, Quebec, Hamilton, Shawinigan, Thunder Bay, Victoria, Windsor

* These descriptions are very generalized: the groups are defined by analysis of eleven principal components.

SOURCE: Based on L. J. King (1966), reprinted in R. L. Gentilcore, *Geographical Approaches to Canadian Problems.*

(4) eight centres that specialize in sectors other than industry and wholesale trades: e.g., Rouyn (mining) and Ottawa (government)
(5) the four main metropolitan centres: manufacturing is more important than wholesale trade in Montreal and Toronto, but not in Vancouver or Winnipeg.[3]

A complete classification should go beyond functional considerations to consider the age of the urban unit, its socioeconomic groups, population density, mobility and age structure, the economic structure, housing conditions, the metropolitan web, relative accessibility and, something distinctively Canadian, the influences exerted by the United States, by external markets, by the cultural mosaic and by the frontier. Using some of these variables, Leslie King subdivided the 106 towns which had more than 100,000 inhabitants in 1961 into nine groups (table 19.3).[4]

3. *Ibid.*

4. L. J. King, "Cross-sectional Analysis of Canadian Urban Dimensions: 1951–1961", *The Canadian Geographer* X (1966): 205–224.

Conclusion

In Canada, so far as absolute and distinctive dominance are concerned, there are many highly specialized towns and cities.

Away from a dense network where cities can provide complementary services to each other, this degree of functional specialization seems a vulnerable system. However, because of the separation of the main market from the rest of the country, the urban units in both western Canada and the Maritimes tend to be more diversified than those in the Windsor-Lévis axis. Urban functions, hence, reflect the present economic system and will be forces of action or inertia in the future structure of the country.

Chapter 20

Intracity and Intercity Linkages

Lack of research and the size of the topic mean that only a brief look at this fundamental aspect of geography can be provided here. The chief concern is with the competitive dominance of Toronto and Montreal and the relationship between urban foci and the rest of the ecumene.

The City as a Focus for Population Movements

Because of their powerful appeal, cities attract immigrants of all sorts. Since World War II, Greater Montreal has absorbed 200,000 new arrivals from abroad, 80 to 90 per cent of all those who have come to Quebec. Such concentration in the largest centres is the norm; in 1961, 63 per cent of all immigrants were living in the metropolitan centres, although the latter contained only 45 per cent of the total population. Of the four main reception centres, Toronto, Montreal, Winnipeg and Hamilton, the first two provide *home* for the majority of newcomers. The New Canadian has become a metropolitan citizen, whereas those who arrived about 1900 found themselves pioneers on the agricultural frontier. In the towns, the immigrant is less "balkanized" than was once the case; income and type of work tend to replace cultural heritage as a factor determining place of residence.

The cities also attract provincial immigrants. From 1956 to 1961, the majority of the more than 1 million workers who moved from one area to another have affected the urban population, although 80 migrants out of every 100 remained in the same province, 10 went to an adjacent province and 10 others to a more distant province. Whereas Toronto has received more intraprovincial workers than extraprovincial migrants, the opposite has been the case in Vancouver and Calgary. In studying the shifts from one province to another, over Canada as a whole and in relation to the metropolitan centres, Y. Kasahara developed an index of attraction of the largest cities in 1961.[1] For every 137 persons drawn to Toronto, Montreal attracted 119 (table 20.1). This urban attraction is nation-wide; all cities, even St. John's (which on the same scale attracts only 13 people), take immigrants from all provinces. In absolute terms, however, the proportions are not

1. Y. Kasahara, "A Profile of Canada's Metropolitan Centres", *Queen's Quarterly* 10 (1963): 303–313.

TABLE 20.1 Origins of Interprovincial Migrants to the Main Metropolitan Areas, 1961

Metropolitan Area	*Total**	*Nfld.*	*P.E.I.*	*N.S.*	*N.B.*	*Que.*	*Ont.*	*Man.*	*Sask.*	*Alta.*	*B.C.*	*Yukon & N.W.T.*
Toronto	137	15	16	15	14	14		14	12	12	14	11
Vancouver	132	12	13	13	11	13	9	15	16	14		16
Montreal	119	13	14	14	15		10	10	9	10	10	14

* These figures indicate the relative pull exerted by the three centres.

SOURCE: Based on Y. Kasahara (1963).

evenly distributed. Toronto and, more especially, Montreal draw the fewest people from the Prairies; the Atlantic Provinces and Ontario are not major source-regions for Vancouver. The other metropolitan areas tend to recruit migrants on a regional scale; for example, Halifax in the Maritimes and Victoria in western Canada.

The cities, therefore, appear to be the destination for all kinds of migrants intent on becoming urbanites. Though the secondary centres mainly attract inhabitants from their own region, the largest centres have a widespread appeal. This pull is sufficient to leapfrog over the intermediate centres. For example, the unemployed resident of Gaspé does not stop in Quebec but goes directly to Montreal; from the Atlantic Provinces, the more English-speaking Toronto attracts larger numbers than does Montreal, despite the greater distance. The cities, therefore, are prime causes of population mobility and they determine the frequency, volume and distance of population movements.

Canada as a place of residence tends to be rather more urban in winter than in summer. Whereas the warm season leads to a daily or longer-term movement out to the country, many rural people spend "winter in town". In the cities, however, the tourist season and summer employment tend to balance the small seasonal loss.

The Urban Hinterland

The attraction to migrants and the urban shadow on agricultural lands are only two of the many urban influences on the landscape. Virtually all the human and economic geography of Canada could be included under this heading. Some very complex, and not necessarily continuous, exchanges develop and are maintained among the urban cores, their periurban fringes, their region, their province and even the whole country.

These relations have a dual character. For example, Joliette, Quebec (20,127 inhabitants), provides its hinterland with educational, religious and professional services, a commercial structure and a little employment. In return, the region provides it with raw materials (tobacco, water), food, customers, and recreation sites (the Laurentians). Between such centres and their regions, relationships are seldom complete, exclusive and uniform. Berthier in the east and l'Assomption and especially Montreal in the west also communicate directly with parts of the Joliette region. Thus, there are spatial thresholds beyond which the relations between the nucleus and the wider region are no longer dominant. Some functions are also lost, in whole or in part, to wider linkages. The town of Joliette has not been able to absorb all the emigration from the three or four neighbouring counties, and the rest have gone elsewhere. Similarly, except in a few fields such as theology, Joliette does not provide education at the university level and the colleges look to Montreal. In regard to certain religious communities, however, Joliette serves the whole province and even the whole of Canada.

The relationships between town and country exist in a variety of links which may be total, dominant, secondary, complementary, occasional or unusual in character. If we alter the criteria used to define these links, we necessarily alter the apparent limits of urban attraction. The limits of urban influence must therefore necessarily remain ill defined, since the boundaries of such dependence are in fact elastic in character and not clear and unchanging lines.

An area of southern Quebec may be taken as an example of the complexity and even the confusion of spheres of influence and urban attraction. Several criteria have been used to identify the varied hinterlands of the town of Sherbrooke (80,457) in the Eastern Townships. Three hemispherical zones are recognized, bounded on the south by the border with the United States and by the Appalachians. A sphere of very close links, about 80 miles (128.7 kilometres) in diameter, extends throughout the counties of Sherbrooke, Stanstead, Compton and Richmond, through much of the counties of Wolfe and Frontenac, and to a lesser extent into the counties of Brome and Shefford. Beyond it, a second zone of less importance about 20 miles (30 kilometres) in width affects eight counties, and a third zone, "the limits of attraction", takes in a dozen. These three zones are composed of areas that, in terms of each individual item, have very different limits. Take, for example, the main zone. Although a radius of 15 miles (24 kilometres) contains the area within which daily commuting takes place, and although Granby provides a nearby western limit to Sherbrooke's financial influence, the commercial influence of "the Queen of the East" extends for at least 70 miles (112 kilometres). There is no longer any equivalence between the boundaries of the diocese, high school zones, media circulation areas, hospital services or administrative units. Still less do such limits coincide in the zones of secondary influence.[2] This absence of clear definition in urban spheres of influence suggests that hinterlands overlap.

Seen on the small scale, marginal areas either benefit from this competition or are scarcely served at all. In southern Quebec, the metropolitan influences of Montreal, Quebec, Sherbrooke and Trois-Rivières do not link up completely and leave some marginal areas with poor services (see fig. 20.1); the latter are forced to rely on secondary centres such as Thetford and Granby.[3]

These conflicts of power are found in all Canadian regions. Take the Atlantic Provinces as an example. St. John's and Halifax dominate their respective provinces, and Corner Brook (Newfoundland) and Sydney (Nova Scotia) have only regional importance. The influence of Moncton overflows from eastern New Brunswick to affect the much less urbanized Prince Edward Island.

In western Canada, Vancouver dominates most of the interior of British Columbia, as Winnipeg used to dominate the Prairie Provinces before that dominance was checked by the recent tremendous growth of Edmonton and Calgary.

Very special hinterlands of metropolitan areas are the "longitudinal" Norths shown in fig. 9.3. The Atlantic North is almost entirely within the sphere of influence of Montreal,

---

2. P. Cazalis, "Sherbrooke: sa place dans la vie de relations des Cantons de l'est", *Cahiers de Géographie de Québec*, No. 16, 1964, pp. 165–197.
3. [Quebec], "Zones d'influence des pôles d'attraction du Québec, 1961–1964", *Quebec Yearbook 1966–1967*, reports of the Bureau of Economic Research (Quebec, 1967).

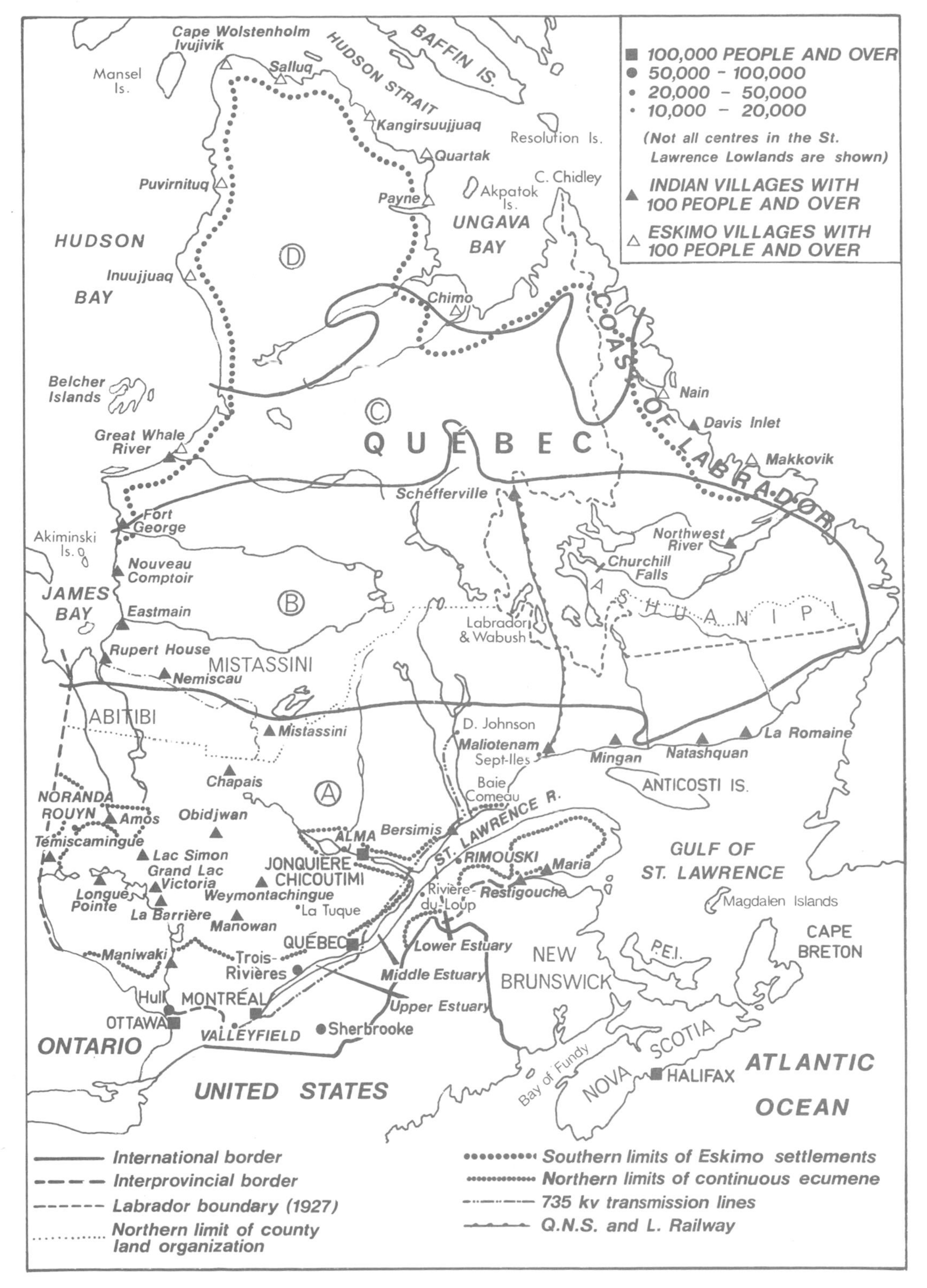

Fig. 20.1 Urban centres of Quebec and Labrador. (After H. Dorion, *Atlas du monde contemporain.*)

Ottawa and Toronto; Quebec City is concerned with New Quebec and St. John's with Labrador. In western Canada Edmonton, Yellowknife and Ottawa are the effective capitals of the Mackenzie North. Such examples, however, raise the more general question of links or competition between individual cities.

The Urban Hierarchy

If the population of different cities is compared, a pattern of spatial dominance or primacy begins to emerge. For example, by comparing the populations of the two largest cities in each province, three types of situation clearly exist. In Quebec, Manitoba and British Columbia, there is overwhelming dominance by a single city. The second cities, Quebec, Brandon and Victoria, account for less than 20 per cent of the population of Montreal, Winnipeg and Vancouver. In these "single-headed" provinces, the entire urban system is likely to be poorly structured, especially since each of the main cities is in a peripheral position. If metropolitan rather than municipal populations were used, Ontario would also belong to this first group. By contrast, forming a second category are the "twin-headed" provinces; Calgary has 92 per cent of the population of Edmonton, Saskatoon 90 per cent that of Regina and Moncton 79 per cent that of Saint John. In Alberta, Saskatchewan and New Brunswick, therefore, the population balance suggests a division of provincial structures and influence, and even a separation of external links. This type of intraprovincial arrangement also seems to have its problems. Lastly, there seems to be a better relationship between the populations of the main cities in Ontario, Nova Scotia, Prince Edward Island and Newfoundland, where the second city has, respectively, 52 per cent, 50 per cent, 47 per cent, and 39 per cent of the population of the largest centre. In terms of population, therefore, only a minority of provinces exhibit a regular hierarchy. The real hierarchy, however, depends mainly on organic intercity linkages which do not necessarily reflect population sizes.

The order of precedence of urban areas depends on a multitude of factors, some of which are very general in character. Winnipeg, the only metropolis in the Prairies before 1920, has tended to lose part of its traditional supraprovincial function as commerce has gravitated westwards and the cities of Alberta have grown in size. Conversely, Vancouver has benefited from improvements in communications to extend its dominance in the B.C. interior.[4] Some structural elements influence the character of the individual links in the urban chain. In central Saguenay, for example, the nature of the conurbation predisposes it more to a competitive separation among its component centres than to their interdependence; complementary industrial relationships encourage the integration of some Ontario cities. In Quebec, history has left a badly balanced urban network. In the west is the enormous bulk of Montreal; in the east is Rimouski, which has some difficulty in being recognized as a regional capital. There is a significant lack of linkages in the provincial urban hierarchy. The prestige of Montreal, the fragmentation and elongation of the main area of ecumene, and a lack of planning have all led to a very centralized system without the balance provided by towns of intermediate size; this lack creates problems at the local level. In the Canadian North there are special problems; distance, local needs, and political decisions have led to the creation of small competing centres. The four administrative and service towns of Fort Smith, Hay River, Yellowknife and Inuvik do not form an integrated system. In this sporadic ecumene, the relationship between the settled sites and the barren areas must obviously differ from the organic links between town and country in southern Canada.

4. J. L. Robinson and W. G. Hardwick, "The Canadian Cordillera", in J. Warkentin, *Canada. A Geographical Interpretation*, pp. 438–472.

Plate 20.1 The interchange between Highways 400 and 401 in Toronto. Opposition to freeway building is growing, but the situation of Toronto, on the shore of Lake Ontario, makes an east-west freeway (401) for through traffic north of the city essential. Similarly, the main highway north to Barrie (400) requires this complex interchange with 401. The main question in the 1970s is not "Did we need freeways like 400 and 401?", but "Now that we have these, do we need more?".

Primacy Compared: Montreal and Toronto

The most notable feature of the urban hierarchy is the similar position held by Montreal and Toronto.[5] In Canada such metropolitan competition is not unique; besides Montreal and Toronto there are Edmonton and Calgary in Alberta, Saskatoon and Regina in Saskatchewan, Saint John and Moncton in New Brunswick. There are also other similar balances between towns in every size of region.

As an extension of the differences in population between one city and another, the present author has devised an index of urban primacy (*IP*).[6] Based on more than twenty demographic

5. D. P. Kerr, "Metropolitan Dominance in Canada", in J. Warkentin, *Canada. A Geographical Interpretation*, pp. 531–556.
See also P. Dagenais, "La métropole du Canada: Montréal ou Toronto?", *Revue de Géographie de Montréal* 23 (1969): 27–37.

6. L.-E. Hamelin, G. Cayouette and R. de Koninck, "Un indice de primatie appliqué à la concurrence entre Montréal et Toronto", *Revue de Géographie de Montréal* XXI (1967): 389–396.

TABLE 20.2 Measures of Primacy: Montreal and Toronto, 1970

Criteria	*Montreal*		*Toronto*	
	Relative Value	*Weighted Dominance*	*Relative Value*	*Weighted Dominance*
Population:				
(1) Similarity to ethnic composition of Canada	100		120	1
(2) Attraction to inter-provincial migrants	100		115	1
(3) Total population	104	1	100	
(4) Bilingual proportion of population	990	3	100	
(5) University students	180 (?)	1	100	
(6) Percentage of provincial population	138	1	100	
(7) Dominance as compared to second city in same province	125	1	100	
(8) Labour force	102	1	100	
	1,839	8	835	2
Economics:				
(9) Incomes	100		104	1
(10) Value of retail sales	100		104	1
(11) Family expenditures	100		112	1
(12) Bank accounts	100		139	1
(13) Stock exchange volume	100		258	2
(14) Number of head offices of mining companies	100		700	3
(15) Port traffic	418	2	100	
(16) Percentage of value of industrial output	100		110	1
	1,118	2	1,627	10
Miscellaneous:				
(17) Stanley Cups since 1935	110	1	100	
(18) Number of hours of television produced	117	1	100	
(19) Sale of daily and weekly newspapers	145	1	100	
(20) Proximity to New York	105	1	100	
(21) Hotel bedrooms	250 (?)	2	100	
(22) Book publication	100		220 (?)	2
(23) Airline passengers	100		131	1
(24) Ratio of cars to population	100		105	1
	1,027	6	956	4
Total	3,984	16	3,418	16

and economic criteria, these new formulae allow the relative importance of metropolitan areas to be more precisely quantified, especially those which are close together.

Because the data used in applying one criterion cannot be directly compared with a different set of data used in another criterion, each criterion is allotted a minimum value of 100 and the maximum value is then calculated in terms of this base (table 20.2). For example, if the 2,628,043 inhabitants of Greater Toronto are considered as 100, the 2,743,208 inhabitants of Greater Montreal represent 104.

If calculations are to be kept simple, there are two choices. The first is to sum up the values of each individual criterion, which gives 3,984 units to Montreal and 3,418 units to Toronto. This method is rejected because of some wide differences in certain criteria that distort the overall picture. For example, the ratio of head offices of mining corporations favours Toronto, while the bilingual criterion benefits Montreal even more. A better method is one that allows a certain amount of weighting of criteria. Three classes have been distinguished: a ratio of less than 2; ratios between 2 and 5 (e.g., for stock exchange activity: Montreal, 100; Toronto, 258); and a ratio over 5. A value of 1 is given to examples in the first class, a value of 2 to those in the second and of 3 to those in the third group. Hence, for example, a value of 1 in favour of Montreal is allotted in respect to total population.

The sum (D) of the individual criteria may be expressed by dividing the higher total by the lower; hence, in this case,

$$IP = \frac{Da\text{ (Montreal)}}{Db\text{ (Toronto)}}$$

where IP is the index of primacy.

Although Montreal has a larger population than Toronto, its overall primacy is very debatable. The primacy index, IP or 16/16, gives a ratio of 1.00; that is to say, no dominance of one city over the other. Further analysis of the evidence clearly shows small sectorial differences between the two cities. The relative values in the first class of dominance (ratio less than 2) give 1,126 units to Montreal and 1,040 to Toronto. In the second and third classes, Toronto is on an equal footing with Montreal. The two are hence close to equivalent in importance.

Apart from some very specialized characteristics like the headquarters of mining corporations, bilingualism and port traffic, dominance in individual characteristics is very slight. For eighteen of the twenty-five criteria (i.e., 75 per cent of the cases), relative predominance falls only in our first class.

Fig. 20.2 clearly demonstrates that the overall dominance is not made up of the same elements and that it reflects the specific characteristics of the two metropolitan areas themselves. Montreal, with eight dominant criteria out of ten, clearly leads in regard to population, while Toronto (ten out of twelve) is as significant in terms of the economy.

In Canada as a whole, there is no single metropolis that dominates the entire country. The influence of Winnipeg or even that of Vancouver scarcely reaches eastern Canada. The Atlantic region lacks any supraprovincial centre. Moreover, although the border cities are normally part of the sphere of influence of U.S. cities, such as Windsor (Detroit), Niagara (Buffalo), Fort Frances (Minneapolis), and Saint John (Boston), the Canadian urban network does not form part of a large continental pattern. Canada is a giant with two heads, each tending to complement the other. Montreal is undoubtedly the cultural metropolis of all French Canada, as Toronto is of English Canada. The English-speaking culture of Montreal (McGill University) is certainly to be reckoned with, whereas Toronto's cultural influence is very limited in French Canada. On the economic scene, Toronto is the metropolis

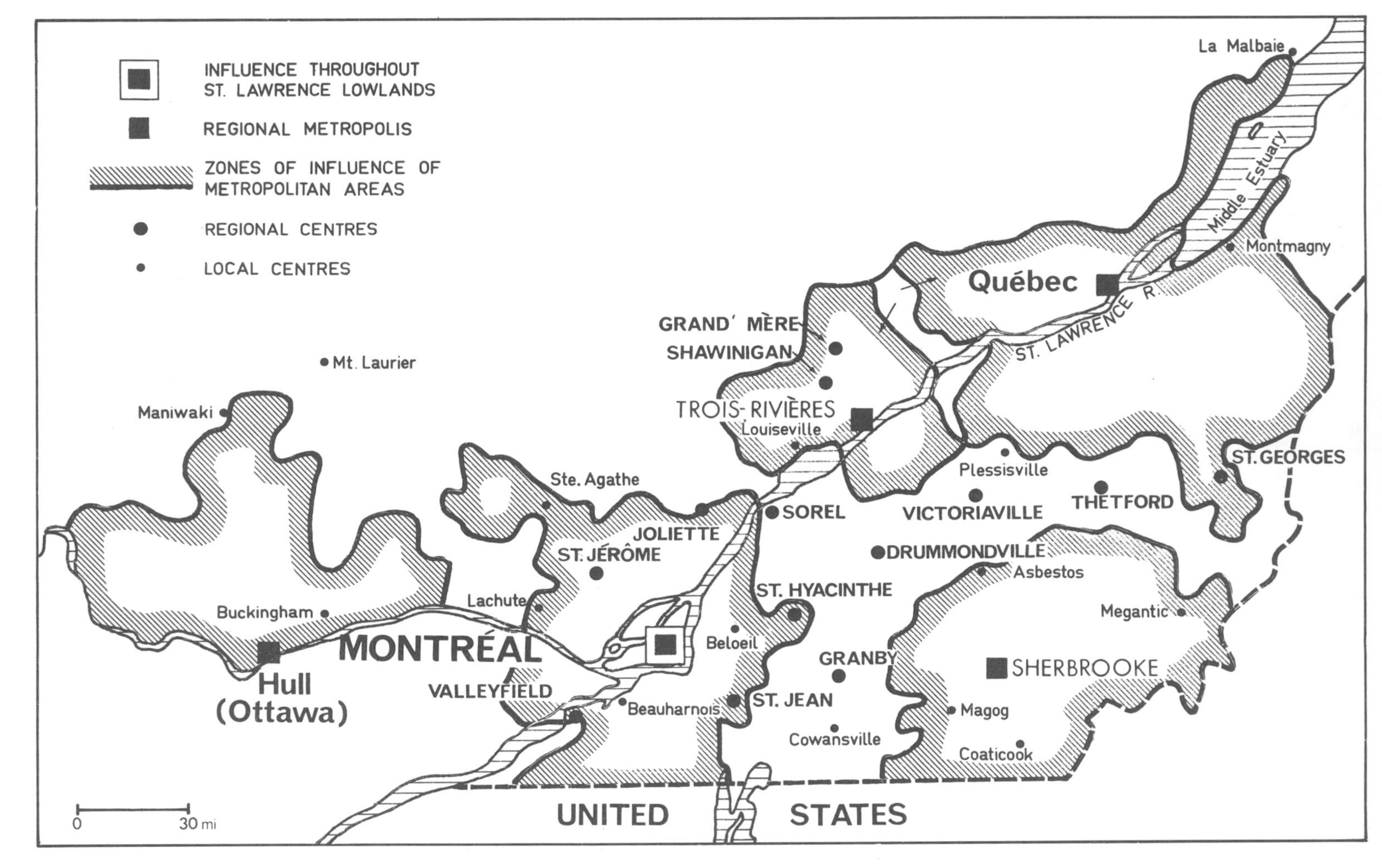

Fig. 20.2 Metropolitan influences in southern Quebec. (Data from the Economic Research Bureau.)

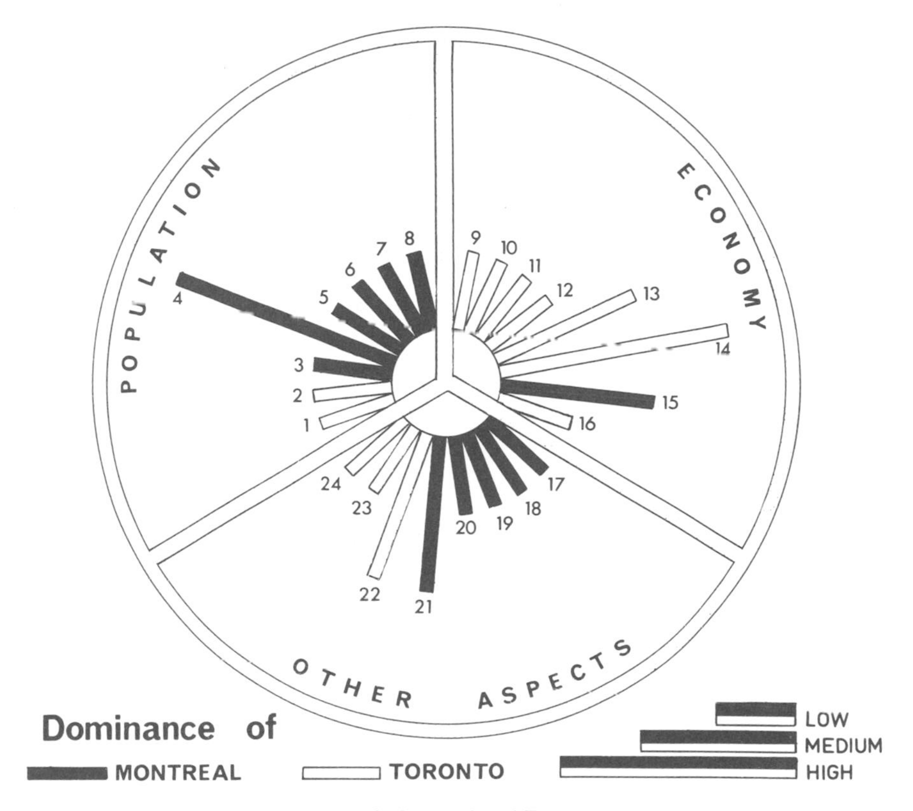

Fig. 20.3 Sectorial dominances of Montreal and Toronto.

of English Canada. The economic influence of Toronto penetrates French Canada and that of Montreal similarly affects English Canada. The slight demographic primacy of Montreal over Toronto does not automatically make it the uncontestable metropolitan centre of Canada. Metropolitan function should be due more to functional factors than to population size. Montreal and Toronto, therefore, have different as well as competing influences.

Canada has two major cities just as it has two main cultural groups. Because of the economic activity exercised by its English-speaking community, the influence of Montreal on Canada is greater than the relative size of French Canada implies, but the second largest French-speaking city in the world would find it difficult to become the primary city in a country which is 70 per cent English-speaking. On the other hand, the Pan-Canadian role of Toronto is equally limited; Toronto is too closely related to the English-language culture to be a metropolis for the 4 million Canadians who speak only French. This division is, in any case, far from solely linguistic. It concerns also the hinterland, the mentality, and the general culture; it is here to stay.

Given such rivalry, it is reasonable to anticipate that the competition for primacy which has existed for almost a century will remain a permanent characteristic.[7] Since neither city is the capital of Canada, national administrative means cannot be relied upon to give a definite advantage to one over the other. Any major change in the political structure of Canada will modify the nature of this competition for metropolitan leadership.

The absence of a single Canadian metropolis and the continuing rivalry between Montreal and Toronto are not the only characteristics of the urban structure of Canada. The majority of regions suffer from a lack of intermediate units and from an inadequate intraregional structure. In the provision of services in rural areas, the units that serve a wider area, the village that serves only a single parish and the dependent hamlet can be identified. Because of the low densities of population and the level of income, the distribution of these facilities is scattered and unorganized. Haphazard urban development has not resulted in a perfectly balanced hierarchy.

7. D. C. Masters, "Toronto versus Montreal, Struggle for Financial Hegemony 1860–1875", *Canadian Historical Review*, No. 22, 1941, pp. 131–146.

Conclusion

Through its importance in the last thirty years, the urban world, in terms of influence, facilities, output and residence, has shaped Canada. This is all the more reason why a superagency of urban affairs that could tackle the major urban problems is necessary. These problems include municipal revenues; suburban growth; intra-city transport (a problem involving both ease and cost); the balance of foreign and internal migration with local employment possibilities; the redevelopment of town centres; the restoration and maintenance of a harmony in the function, space and structure of urban areas; legislation to provide an effective overall authority for the metropolitan regions; the role of growth centres in regional development; and the reorganization of urban patterns within provinces and the major regions of Canada. In 1971, the government of Canada created a Ministry of State for Urban Affairs to be responsible for the development of federal urban policy.

Conclusion

A Young and Different Canadian Identity

What is it about Canada that the Quebec poet Anne Hébert calls "the resemblance of the heart to the original land"? In attempting the difficult evaluation of the development of a Canadian unity, we shall be concerned here only with mental attitudes. We have already looked at the distinctiveness of Canada's landscape, political structures, economic life and people. The notion of belonging to a Canadian community can be looked at from outside as well as from within.

As compared to others, the Euro-Canadian is no longer the stranger that his ancestor was and one can see in this a proof, though a rather negative one, of a certain degree of intimacy with his new country. A French Canadian is recognizably different from the French of France; similarly, there are few Canadians of British descent who would still pass as European British. The majority of other immigrants define themselves more by what they have become than by what they were in their homeland. A Canadian is no longer to be confused with a citizen from the United States, even though both have similar North American habits. In other words, the problem of the Canadian identity does not arise because of insufficient differences from other people.

Seen from within, Canada seems to lack the complete or general unity which characterizes the mentality of countries with a strong national core. Even if we exclude the 2 million foreigners who, though contributing to the importance of Canada by their presence and their economic significance, are not sociologically Canadian, the other Canadians do not demonstrate any basic homogeneity. Hugh MacLennan has accurately noted that Ottawa does not so much assume the role of a real capital as it leaves Toronto and Montreal competing for primacy. To say the least, a cultural gap has never ceased to divide the country to some degree; in 1961, only 8 per cent of French-speaking husbands had an English-speaking wife and only 5 per cent of English-speaking men had a French-speaking wife, Catholic opposition having stood in the way of mixed marriages. Between English Canadians and French Canadians, contacts are small and it has been predicted that the two communities will never be one. The ethnic diversity has led, in its turn, to a plurality of cultural systems: a "national" type (in fact, English Canadian); a French-Canadian culture mainly rooted in Quebec; the folklore contributions of other cultural groups; and the universal North American tinge. Both the juxtaposition and the regional character of these cultures mean that Canada does not have those characteristics (newspapers, religions and educational systems) that, in a country of greater homogeneity and stronger centrality, would clearly be recognizable as "national". There is nothing in the way of a single language spoken by everyone; nor do the press agencies prepare the same sets of news for each of the two language networks.

The following are recommended for their discussions of Canadian identity:

(1) J. B. Brebner, "Nationality in Diversity", in *Canada. A Modern History* (Ann Arbor: University of Michigan Press, 1960), pp. 515–526;
(2) A. H. Clark, "Geographical Diversity and the Personality of Canada", in *Land and Livelihood* (Christchurch, 1962), pp. 23–47;
(3) W. L. Morton, *The Canadian Identity* (Toronto: University of Toronto Press, 1967).

The standard list of supposedly characteristic features is misleading because the items are scarcely Canadian. Canada thus seems to be more a land and a political state than a culture, especially a unified culture. On the national scale, there has been a delay in developing a collective will in regard to political and economic structures such as the Confederation of 1867.

The main elements in this brief and heterogeneous national identity have often been pointed out: the vastness of the country; the length and fragmented character of southern Canada; the deep-rooted opposition that the two founding groups accumulated in Europe and maintained in their colonial rivalries; the spontaneous social segregation of the Indians; the political, economic, demographic, trade union and cultural impact of the United States; the multicultural immigration which has not been subjected to a melting-pot policy; the lack of historic occasions which build a clear national identity; and, lastly, deep political divisions (e.g., over the powers of the Assembly of Lower Canada, the Riel rebellions from 1870 to 1885, and nowadays, the fiscal confrontation between Ottawa and some provinces). All these characteristics and events cannot lead to an easy flowering of real feelings of national identity. Up to now, neither nature nor history has fostered the full development of a Pan-Canadian culture.

On the other hand, however, people and things have encouraged some currents of integration. The fact that the whole of Canada was opened up via the St. Lawrence has led not merely to an undeniable priority to be given to Ontario and to Quebec, but also to the establishment of a federal structure. In Canada, imperial ambitions and the affirmation of the British fact against the United States have tended to work in favour of centralization. Nothing, however, has been more important in creating a certain Canadianization than the transcontinental enterprises, especially the railway building of the last century. Today the multiple, subtle and complex elements of Canadianism mainly comprise the transport and telecommunications systems, the dominance of the English language, the centralizing policy of the federal government, Canadian citizenship, the currency, the postal system, the major political parties, the chains of stores and of newspapers, the financial systems, national television, interprovincial movements and even the discussion of certain problems of universal interest, such as the division of powers among governments. Such manifestations of Canadianization are undoubtedly contributing to the development of some form of "national" behaviour.

The time dimension should not be forgotten; a few centuries is a very brief period in which to acquire homogeneous characteristics, at least those that are well defined. In theory, the weak Canadian identity could be due as much to the limited period of the experiment as to the intrinsic difficulties of the objective. Further, the nature of the Canadian identity, taking cultural features alone, has been greatly modified in the course of history. Around 1800, to be Canadian was to be French Canadian; half a century or so later, the English Canadians born in Canada called themselves "Native Canadians". Today, for a strong minority, the Canadian identity is not defined as an extension of the two founding peoples. The qualification *Canadian* therefore describes and joins together very complex elements.

In cultural terms, Canada has all sorts of inequalities, from simple local variations to differences in kind, and even sharp discontinuities. The Canadian identity is much less a unity that must be sought than a plurality that should be recognized. For Canada the only way to be one is to accommodate several; "national" feelings depend not on a reduction to one indefinite whole, but on the optimal reconciliation of differences. The fact that "Canadianism makes a deliberate choice in favour of diversity"

(L. B. Pearson) is almost a paradox because it assumes that a great number of people will take a reasonable attitude on an emotional issue.

The Internal and External Dominance of English

In a variety of ways, an English-speaking dominance is apparent in Canada from both outside and within the country. Since the internal dominance of English attitudes in the various demographic, linguistic, economic and political spheres has already been emphasized, only external relations need be discussed here. Two countries are relevant.[1]

Great Britain

At the time of Confederation, Canada was politically British and was not granted direct responsibilities in regard to external relations; the federation of the colonies remained in the British Empire. The slow development which occurred subsequently was marked by contradictory tendencies but led in the end to the official independence of Canada in 1931. Recent years have been marked by a swifter dismantling of British connections: for example, at the end of World War II, the Canadianization of former English capital, the abolition of appeals to the Privy Council in 1949, and the nomination of the first Canadian governor-general in 1952. Recruiting members for such very British Canadian institutions as the Sons of England, the Daughters of England and The Imperial Order Daughters of the Empire has become difficult. Britishness is not dead in Canada, however. In 1962, the (Conservative) government in Canada was firmly opposed to the entry of Great Britain into the Common Market. In 1964, after ninety-seven years of Canadian confederation, the federal government had considerable difficulty in getting a distinctively Canadian flag adopted. It was Prince Philip who represented Canada at the fiftieth anniversary of the Battle of Vimy in France; in 1973, Queen Elizabeth is scheduled to visit some English-speaking provinces. All these little symbols of a colonial nature obviously irritate Canadian nationalists (especially French Canadians), to which the beguiling reply is that these thin links with Great Britain serve precisely as a defence against the influence of the United States. But a British outlook no longer plays the protecting role in this respect that it did in the nineteenth and early twentieth centuries.

The Commonwealth has become very different from that of the imperial age, when Great Britain easily exercised her leadership in pursuit of objectives which appeared common to all its members. Nowadays, Canada as the senior member may be observed occasionally trying to reconcile opposing points of view between London and some Third World states; similarly, Canada today discusses various problems directly with Great Britain outside the framework of the Commonwealth. But the most serious problem concerns the entry of Great Britain into the European Common Market, voted by London in July 1972. The traditional commercial link between Canada and Britain is in danger of being diminished at precisely the moment when it seems desirable to maintain an important level of trade with partners other than the too powerful United States. Such a desirable alternative market might one day be a United Europe.

In any event, as a result of developments outside as well as inside Canada, the United Kingdom could at least become reconciled to the idea of a Canadian republic. A political separation within Canada would be a much more radical development. Located between two very unequal parts of Canada—the Atlantic Provinces, and Ontario and western Canada

---

1. H. A. Innis, "Great Britain, The United States and Canada", in M. Q. Innis, ed., *Essays in Canadian Economic History*, pp. 394–412.

—which are almost British in regard to the monarchy but which are strongly influenced by the United States, an independent Quebec would try to reach beyond this twin Anglophone world. Such an extreme situation would force Great Britain into a basic re-evaluation of her relations with North America.

The United States

The external relations of Canada are inconceivable without the varied and omnipresent impact of the United States, the powerful neighbour of Manifest Destiny. New France carried on a lively commercial and military rivalry with the thirteen colonies. In the east and centre of North America, after a very short period of universal British allegiance between 1760 and 1776, the American Revolution poured thousands of Loyalists into the Maritimes, Ontario and Quebec; by refusing to join the Republic they helped to strengthen British influence in Canada. On this occasion, politically as well as mentally, Canada asserted itself against the United States. Since, for about a century (1776, 1812, 1866), the U.S.A. directly threatened Canada's independence, a defensive attitude developed in Canada, even in Quebec. However, for almost a hundred years a peaceful state of affairs has removed the need for frontier fortifications between the two countries.

In time of war, the real or imagined needs of continental protection have forced Canada to accept the emergency policies of the United States. Networks of airports and military camps around 1942 and then radar lines around 1955 have been controlled on Canadian soil by the U.S.A., at first exclusively, and then, in theory, with shared responsibility. Influenced, without any doubt, by her southern neighbour, Canada signed the defence agreement of the North Atlantic Treaty Organization in 1949, an agreement that one of the partners, France, called into question in recent years. NORAD was established at Colorado Springs for the joint air defence of all of North America. These military arrangements still exert a clear influence on Canada's foreign policy.

In several respects, economics have directly influenced policy. A country like Canada, with a strong capacity for production but lacking, for a long time, industrial development and a large internal market, must consider international trade a necessity. In 1926, exports represented approximately 32 per cent of the gross national income. This situation has led to some contradictory political repercussions. On the one hand, these trade relations can lead to diplomatic initiatives extending as far as continental China, with which Canada has conducted a profitable trade in cereals. In this matter, Canada moved away from and ahead of United States policy by wanting Red China in the UN, in accordance with Great Britain and France. Cuba and the U.S.S.R. have also been the objects of rather unconventional attitudes. It thus appears that the need for a freer economic system has encouraged the practice of a certain degree of political freedom. On the other hand, commercial, technological, cultural, trade union and financial links with the United States lead to a definite economic alignment and consequently discourage the creation of an entirely autonomous Canadian foreign policy.

Canada now and then reacts in a way that merits the term *nationalist*. Some years ago the federal government preferred a Canadian route for a pipeline between the Prairies and the St. Lawrence, even though it involved a higher cost of construction, more difficult financing and a more restricted market, all meaning a higher final price for the transported product. Now, many Canadians continue to stand for a moderate nationalism, in order to decrease the foreign role in the resources, industry and trade of Canada, and thus to prevent the United States from controlling the Canadian economy.

Between now and the year 2000, the percentage of foreign capital in certain fields should fall by more than 50 per cent to around 30 per cent. But the future is not necessarily a choice between Canadian autonomy or absorption by the United States; a third way might be a joint agreement on both sides. Some people have seen in the binational agreements of the St. Lawrence in 1959 and the Columbia agreement in 1964, as well as the auto pact of 1965, and perhaps a northern pipeline, the beginnings of a form of common market. Whatever comes of these views, the "dangerous liaison" is likely to remain a continuing factor.

Canada is an international power in terms of output, trade, standard of living, population movements and communications. However, it is not wholly free to act independently of the two reference points, Great Britain and the United States. Canada has never operated outside the magnetic fields of those who are stronger than itself. External political maturity can only be established in stages and by compromise. The slow process of disengagement from these two dominant forces will probably accelerate. It is probable also that in future, France, South America, eastern Europe, Africa, the U.S.S.R. (itself a neighbour) and Japan will be of more importance in Canadian foreign relations.

The Development of a Francophone Quebec

In the presentation of successive French Canadas (chapter 12), emphasis was placed on the drive that has characterized French-speaking Quebec for the last ten years or more in the spheres of culture, economy and even politics, a drive which curiously has not been accompanied by the continuation of a high birthrate. It is an accelerating phase of history that General de Gaulle found "wholly natural", as he himself declared. The French Canadians of Quebec who want to expand in this way are questioning all the forms of domination to which they had appeared accustomed; they have some very definite objectives. English Canadians must accept that the separatists are not the only ones to desire a better situation for French Canada; this is a general attitude. Though all French Canadians agree that French-speaking Quebec must be developed, there are profound differences concerning the political means and the speed of achievement of these objectives: for example, Québécois are divided on the matter of using the Confederation formula regarding the care of French Canadians outside Quebec and the number of official languages in Quebec. If Ottawa, which gets the blame (though, on all the evidence, the inferiority of French Canadians is due to other major causes also), does not succeed in changing the tide, Quebec, whose population is young and has the vote at eighteen years of age, could soon opt for independence by a majority at the provincial government level or at a special referendum. If this situation came to pass, Quebec would have to be ready to assume her responsibilities. The success of such an option would demand much imagination, good will and work, as well as outside assistance. What is impossible to forecast is the real reaction of the powerful non-French-speaking minority of Montreal and that of English Canada both west and east.[2]

The development of Quebec is also related to external factors.[3] After 1760, the political relationships between Canada and the French cultural world were so weak by comparison with English-speaking contacts that it was

2. A. F. Burghardt, "Quebec separatism and the future of Canada", in L. Gentilcore, ed., *Geographical Approaches to Canadian Problems*, pp. 229–235.
3. J. Hamelin, "Quebec and the Outside World, 1867–1967", in *Quebec Yearbook 1968–1969*, pp. 37–60.

unnecessary to look in this direction for an expression of the bicultural character of the country. Not until 1855, almost a century after the battle on the Plains of Abraham, did France re-establish direct relations with Canada.[4] The latter, strongly British until recently, adopted English as its diplomatic language, even in Paris. Imperial preference, the low volume of Franco-Canadian trade, the small number of French immigrants to Canada, the dominant colonial concern of France with Asia and Africa, and a lack of interest on the part of a French Canada which regarded France as "irreligious and blasphemous" all discouraged political contacts between Canada and one of her mother countries. This being so, French Canada lacked an international outlook of its own during the two centuries which followed 1763. It did, however, influence Canadian foreign policy in regard both to Britain and the U.S.A. Quebec, with few emotional reasons for being sympathetic to British attitudes, has often argued that Canada should detach itself progressively from Great Britain, a mental evolution which for the English Canadian came only much later. In regard to the United States, the presence of French Canada has given real substance to the claim to a Canadian personality.

Much more novel, significant and bold have been the events since 1960. Exploiting the internal division of powers between the federal and provincial levels, Quebec entered into the special area of external affairs: on 27 February 1965, Quebec signed an agreement with France; the latter decided, in 1967, to increase its assistance to Quebec tenfold. France's intervention was welcomed by some and frowned upon by others. The outward influence of Quebec is not limited to the "French hexagon" but extends also to the French-speaking countries of Africa, which, through the intermediary of the Department of External Affairs in Ottawa, receive educational assistance; the University of Rouanda has been directed by a Canadian. Even without independence, Quebec is becoming an international power in matters within her competence. Some would argue that Quebec is in a better position than Ottawa to maintain Canadian relations with the French-speaking world; it is doubtful, however, that the federal government, even reluctantly, would entrust Quebec with all of this task. French-speaking Quebec, after having been the first to assume a real Canadian identity back in the eighteenth century, is compelling Canada to redefine itself.

It is in these forms that the Canadian spirit appears to us. Although present-day Canada is not fragmented to the point of constituting a complete mixture like India, nor is it two factions clearly separated as in Malaysia (both of them Commonwealth countries), neither is it a country with a recognized common denominator, such as France. Like numerous other young countries, it is searching for an identity; in the meantime, a real Canadian feeling is beginning to affect the majority of the country's inhabitants.

---

4. P. Savard, *Le consulat général de France à Québec et à Montréal de 1859 à 1914*, Cahiers de l'Institut d'histoire, No. 15 (Quebec, 1971).

Bibliography and Selected Readings

Books and Articles

Andrews, J. T., *A Geomorphological Study of Post-Glacial Uplift with Particular Reference to Arctic Canada*, Institute of British Geographers special publication, No. 2. London: Alden Press, 1970. (156 pp.)

Balikci, A., *The Netsilik Eskimo*, with a foreword by Margaret Mead. Garden City: The Natural History Press, 1970. (264 pp.)

Baulig, H., *Amérique septentrionale. Généralités. Canada.* Paris: Colin, 1936. (315 pp.)

Beals, C. S., ed., *Science, History and Hudson Bay.* 2 vols. Ottawa: Queen's Printer, 1968. (1,057 pp. total.)

Beaulieu, A., Bonenfant, J. C. and Hamelin, J., *Répertoire des publications gouvernementales du Québec de 1867 à 1964.* Quebec: Editeur officiel du Québec, 1968. (544 pp.)

Belanger, M., Brunet, Y., Gauthier, D. and Manseau, H., "Le complexe périmétropolitain montréalais: une analyse de l'évolution des populations totales", *Revue de Géographie de Montréal* 26 (1972): 241–249.

Bertin, L., *Target 2067, Canada's Second Century.* Toronto: Macmillan, 1968. (297 pp.)

Biays, P., *Les marges de l'oekoumène dans l'est du Canada.* Quebec: Laval University Press, for the Centre d'Études Nordiques, 1964. (760 pp.)

Bird, J. B., *The Physiography of Arctic Canada with Special Reference to the Area South of Parry Channel.* Baltimore: Johns Hopkins Press, 1967. (336 pp.)

———, *The Natural Landscapes of Canada. A Study in Regional Earth Science*, with an introduction by F. K. Hare. Toronto: Wiley Publishers of Canada, 1972. (191 pp.)

Blanchard, R., *Montréal et sa région.* Montreal: Beauchemin, 1953. (401 pp.)

Bostock, H. S., *A Provisional Physiographic Map of Canada.* Ottawa: Queen's Printer, 1965. (24 pp.)

Brouillette, B., *Les industries manufacturières du Canada.* Montreal: Institut d'économie appliquée, 1965. (181 pp.)

Brown, R. J. E., *Permafrost in Canada: Its Influence on Northern Development.* Toronto: University of Toronto Press, 1970. (234 pp.)

Camu, P., Weeks, E. P. and Sametz, Z. W., *Economic Geography of Canada.* 1st ed. Toronto: Macmillan, 1964. (393 pp.)

Canada, *Census of Canada.* Ottawa: Information Canada, decennially from 1871. (Bilingual.)

———, *Resources for Tomorrow.* 3 vols. Ottawa: Queen's Printer, 1961. (641 pp., 459 pp., and 566 pp. respectively; bilingual.)

Canada, *Resources for Tomorrow*. Supplementary volume. Ottawa: Queen's Printer, 1962. (225 pp.; bilingual.)

———, *Final Report of the Royal Commission on Canada's Economic Prospects*. Ottawa: Queen's Printer, 1963. (509 pp.; bilingual.)

———, *Annual Report, Economic Council of Canada*. Ottawa: Information Canada, annually from 1964. (Bilingual.)

———, *Canada, One Century 1867–1967*. Ottawa: Queen's Printer, 1967. (504 pp.; bilingual.)

———, *Report of the Royal Commission on Bilingualism and Biculturalism*. 6 vols. Ottawa: Queen's Printer, 1967–1970. (Bilingual.)

———, *Canada North of 60. An Introduction to Resource and Economic Development in the Yukon and the NWT*. Ottawa: Queen's Printer, 1969. (Bilingual.)

———, *Canada Year Book*. Ottawa: Information Canada, annually. (Bilingual.)

———, *Proceedings, Conference on Transport in the Arctic*. 3 vols. Ottawa: Information Canada, 1971. (336 pp., 532 pp., and 409 pp. respectively; bilingual.)

———, *Canada, 1972. The Annual Handbook of Present Conditions and Recent Progress*. Ottawa: Information Canada, 1971. (336 pp.; bilingual.)

———, *List of Theses and Dissertations on Canadian Geography*, Geographical Paper No. 51. Ottawa: Information Canada, 1972. (114 pp.; bilingual.)

———, *Proceedings of the Muskeg Conference*. Ottawa: Information Canada, for the National Research Council, published annually.

Canadian Association of Geographers, "Canadian Geography, 1967", *The Canadian Geographer* XI (1967): 196–371.

Cardinal, H., *The Unjust Society: The Tragedy of Canada's Indians*. Edmonton: Hurtig, 1969. (171 pp.)

Careless, J. M. S. and Craig Brown, R., *The Canadians. 1867–1967*. Toronto: Macmillan, 1967. (856 pp.)

Caves, R. E. and Holton, R. H., *The Canadian Economy: Prospect and Retrospect*. Cambridge: Harvard University Press, 1959. (676 pp.)

Centre d'Études Nordiques, Laval University, *Bibliography of the Quebec-Labrador Peninsula*. 2 vols. Compiled by A. Cooke and F. Caron. Boston: Prentice-Hall, 1968. (432 pp., and 381 pp. respectively; bilingual.)

Chapman, L. J. and Brown, D. M., *The Climate of Canada for Agriculture*. The Canada Land Inventory. Ottawa: Queen's Printer, 1966. (24 pp.)

Chapman, L. J. and Putnam, D. F., *The Physiography of Southern Ontario*. Toronto: University of Toronto Press, 1951. (284 pp.)

Clark, A. H., *Acadia. The Geography of Early Nova Scotia to 1760*. Madison: University of Wisconsin Press, 1968. (450 pp.)

Cook, R., ed., *French-Canadian Nationalism; An Anthology*. Toronto: Macmillan, 1969. (336 pp.)

Creighton, D., *Dominion of the North. A History of Canada*. Toronto: Macmillan, 1966. (619 pp.)

Currie, A. W., *Canadian Transportation Economics*. Toronto: University of Toronto Press, 1967. (719 pp.)

Dansereau, P., *Dimensions of Environmental Quality*. Montreal: Institut d'urbanisme, 1971. (109 pp.)

Deffontaines, P., *L'homme et l'hiver au Canada*. Paris: Gallimard, 1957. (293 pp.)

Dictionary of Canadian Biography. In progress. Toronto: University of Toronto Press. (See Vol. I, *1000–1700*, 1966; Vol. II, *1701–1740*, 1969; and Vol. X, *1871–1880*, 1972.)

Dionne, J. C., *Aspects morpho-sédimentologiques du glaciel, en particulier des côtes du Saint-Laurent*. Report of the Federal Forestry Research Laboratory. Quebec, 1970. (412 pp.)

Dorion, H., *La frontière Québec–Terre-Neuve. Contribution à l'étude systématique des frontières*. Quebec: Laval University Press, 1963. (316 pp.)

Douglas, R. J. W., ed., *Geology and Economic Minerals of Canada*. Ottawa: Information Canada, 1971. (838 pp.; bilingual.)

Elliott, J. L., ed., *Immigrant Groups. Minority Canadians 2*. Toronto: Prentice-Hall, 1971. (215 pp.)

Elton, D. K., ed., *One Prairie Province? Conference Proceedings and Selected Papers*. Lethbridge: Lethbridge Herald, 1970. (455 pp.)

Faucher, A., *Histoire économique et unité canadienne*. Montreal: Fides, 1970. (296 pp.)

Firestone, C. J., *Problems of Economic Growth: Three Essays and Economic Projections for Canada, 1961–1991*. Ottawa: University of Ottawa Press, 1965. (196 pp.)

Fortier, Y. O., Blackadar, R. G. et al., *Geology of the North-Central Part of the Arctic Archipelago, NWT (Operation Franklin)*. 2 vols. GSC Memoir, No. 320. Ottawa: Queen's Printer, 1963.

Forward, C. N., "A Comparison of Waterfront Land Use in Four Canadian Ports: St. John's, Saint John, Halifax, and Victoria", *Economic Geography* 45 (1969): 155–170.

Found, W. C. and McDonald, G. T., "The Spatial Structure of Agriculture in Canada. A Review", *The Canadian Geographer* XVI (1972): 165–180.

Frégault, G., *Canada: The War of the Conquest*, translated by M. C. Cameron. Toronto: Oxford University Press, 1969. (427 pp.)

Gentilcore, R. L., ed., *Canada's Changing Geography*. Toronto: Prentice-Hall, 1967. (224 pp.)

———, *Geographical Approaches to Canadian Problems. A Selection of Readings*. Toronto: Prentice-Hall, 1971. (235 pp.)

Glazebrook, G. P. de T., *A History of Transportation in Canada*. 2 vols. Carleton Library, No. 11 and No. 12. Toronto: McClelland and Stewart, 1964.

———, *Life in Ontario; A Social History*. Toronto: University of Toronto Press, 1968. (316 pp.)

Groulx, L., *La découverte du Canada. Jacques Cartier*. Montreal and Paris: Fides, 1966. (194 pp.)

Hamelin, L.-E. et al., *Canada*. Montreal: ERPI, 1968–1972. (Textbook, colour slides and wall maps for teaching the geography of Canada at the college level.)

Hamelin, L.-E. and Cook, F. A., *Illustrated Glossary of Periglacial Phenomena*. Quebec: Laval University Press, 1967. (237 pp.; bilingual.)

Hare, F. K., *A Photo-Reconnaissance Survey of Labrador-Ungava*, Geographical Branch Memoir, No. 6. Ottawa: Queen's Printer, 1960. (82 pp.)

Hare, F. K., ed., *The Tundra Environment*. Toronto: University of Toronto Press, 1970. (50 pp.)

Hare, F. K. and Hay, J. E., "Anomalies in the Large-Scale Annual Water Balance Over Northern North America", *The Canadian Geographer* XV (1971): 79–94.

Harris, R. C., *The Seigneurial System in Early Canada. A Geographical Study*. Quebec: Laval University Press, 1966. (247 pp.)

Honigman, J. and Honigman, I., *Arctic Townsmen: Ethnic Backgrounds and Modernization*. Ottawa: St. Paul University Press, 1970. (303 pp.)

Inman, M. K. and Anton, F. R., *Economics in a Canadian Setting*. Toronto: Copp Clark, 1965. (666 pp.)

Innis, H. A., *The Fur Trade in Canada*, with a foreword by R. W. Winks. Toronto: University of Toronto Press, 1964. (446 pp.)

Innis, M. Q., ed., *Essays in Canadian Economic History*. Toronto: University of Toronto Press, 1962. (418 pp.)

Institut de Géographie, Laval University, *Mélanges géographiques canadiens offerts à Raoul Blanchard*, with a foreword by L.-E. Hamelin. Quebec: Laval University Press, 1959. (494 pp.)

Irving, R. M., *Readings in Canadian Geography*. Montreal: Holt, Rinehart and Winston, 1968. (398 pp.)

Jackson, C. I., *The Spatial Dimensions of Environmental Management in Canada*, Dept. of Fisheries and Forestry Geographical Paper, No. 46. Ottawa: Information Canada, 1971. (29 pp., bilingual.)

Jenness, D., *The Indians of Canada*. National Museum of Canada Bulletin, No. 65. Ottawa: Queen's Printer, 1960. (452 pp.)

Juillard, E., *L'économie du Canada*. Paris: Presses Universitaires de France, 1968. (128 pp.)

Kalbach, W. E. and McVey, W. W., *The Demographic Bases of Canadian Society*. Toronto: McGraw-Hill, 1971. (354 pp.)

Kendrew, W. G. and Currie, B. W., *The Climate of Central Canada. Manitoba, Saskatchewan, Alberta and the Districts of Mackenzie and Keewatin*. Ottawa: Queen's Printer, 1955. (194 pp.)

Kerr, D. and Spelt, J., *The Changing Face of Toronto*. Geographical Branch Memoir, No. 11. Ottawa: Queen's Printer, 1965. (163 pp.)

Lasalle, P., "Late Quaternary Vegetation and Glacial History in the St. Lawrence Lowlands, Canada", *Leidse geologische Mededelingen* 38 (Leiden, 1966): 91–128.

Legget, R. F., ed., *Soils in Canada: Geological, Pedological and Engineering Studies*, Royal Society of Canada Special Publications Series, No. 3. Toronto: University of Toronto Press, 1961. (224 pp.)

Leim, A. H. and Scott, W. B., *Fishes of the Atlantic Coast of Canada*, Fisheries Research Board Bulletin, No. 155. Ottawa: Queen's Printer, 1966. (485 pp.)

Lévesque, R., *La solution. Le programme du parti québécois*. Montreal: Editions du Jour, 1970. (126 pp.)

Lithwick, N. H., *Urban Canada. Problems and Prospects*. Ottawa: Information Canada, 1970. (236 pp.)

Longley, R. W., *The Climate of Montreal*. Ottawa: Queen's Printer, 1960. (46 pp.)

Lotz, J., *Northern Realities. The Future of Northern Development in Canada*. Toronto: New Press, 1970. (307 pp.)

Lower, A. R. M., *Colony to Nation. A History of Canada*. Toronto: Longmans, 1946. (600 pp.)

Mackay, J. R., *Mackenzie Delta Area*, Geographical Branch Memoir, No. 8. Ottawa: Queen's Printer, 1963. (232 pp.)

Mackintosh, W. A., *The Economic Background of Dominion-Provincial Relations; App. III, Royal Commission Report on Dominion-Provincial Relations*, Carleton Library, No. 13. Toronto: McClelland and Stewart, 1964. (191 pp.)

Manitoba, *Manitoba 1962–1975. Report of the Committee on Manitoba's Economic Future*. Winnipeg, 1963.

Mayer-Oakes, W. J., ed., *Life, Land and Water. Proceedings of the 1966 Conference on Environmental Studies of the Glacial Lake Agassiz Region*. Winnipeg: University of Manitoba Press, 1967. (414 pp.)

McCrossan, R. G. and Glaister, R. P., eds., *Geological History of Western Canada*. Calgary: Alberta Society of Petroleum Geologists, 1966. (232 pp.)

Megil, W. J., ed., *Patterns of Canada*. Toronto: Ryerson, 1966. (278 pp.)

Mid-Canada Development Foundation, *Essays on Mid-Canada. Presented at the First Session*. Toronto: Maclean-Hunter, 1970. (484 pp.)

Morton, W. L., *The Canadian Identity*. Toronto: University of Toronto Press, 1967. (125 pp.)

Neatby, H., *Quebec. The Revolutionary Age, 1760–1791*. The Canadian Centenary Series, No. 6. Toronto: McClelland and Stewart, 1966. (300 pp.)

Nicholson, N. L., *The Boundaries of Canada, Its Provinces and Territories*. Geographical Branch Memoir, No. 2. Ottawa: Queen's Printer, 1954. (142 pp.)

Newfoundland-Labrador, *Report of the Royal Commission on the Economic State and Prospects of Newfoundland and Labrador*. St. John's: Queen's Printer, 1967.

Northwest Territories, *Annual Report of the Commissioner of the Northwest Territories*. Yellowknife and Vancouver, 1970. (128 pp.; bilingual.)

Ormsby, W., *The Emergence of the Federal Concept in Canada, 1939–1945*. Canadian Studies in History and Government, No. 14. Toronto: University of Toronto Press, 1969. (151 pp.)

Orvig, S., ed., *Climates of the Polar Regions*, Vol. 14 of *World Survey of Climatology*. New York: Elsevier, 1971. (375 pp.)

Ouellet, F., *Histoire économique et sociale du Québec, 1760–1850; structures et conjoncture*. Montreal: Fides, 1966. (639 pp.)

Parry, J. T. and Macpherson, J. C., "The St. Faustin–St. Narcisse Moraine and the Champlain Sea", *Revue de Géographie de Montréal* 18 (1964): 235–248.

Pépin, P. Y., *La mise en valeur des ressources naturelles de la région Gaspésie–Rive Sud*. Quebec: Ministry of Industry and Commerce, 1962. (360 pp.)

Phillips, R. A. J., *Canada's North*. Toronto: Macmillan, 1967. (306 pp.)

Plunkett, T. J., *Urban Canada and Its Government: A Study of Municipal Organization*. Toronto: Macmillan, 1968. (178 pp.)

Podoluk, J. R., *Incomes of Canadians*. 1961 Census Monograph. Ottawa: Queen's Printer, 1968. (356 pp.)

Putnam, D. F. and Kerr, D. P., *A Regional Geography of Canada*. Toronto and Vancouver: Dent, 1966. (520 pp.)

Quebec, *Rapport de la commission d'étude sur l'intégrité du territoire du Québec*. 64 vols. (forthcoming). Quebec, 1967–

———, *Annuaire du Québec, Quebec Yearbook*. Quebec: Quebec Official Publisher, annual.

Ray, A. J., "Indian Adaptations to the Forest-Grassland Boundary of Manitoba and Saskatchewan 1650–1821: some Implications for Interregional Migration", *The Canadian Geographer* XVI (1972): 103–118.

Raymond, C. W., McClellan, J. B. and Rayburn, J. A., *Land Utilization in Prince Edward Island*. Geographical Branch Memoir, No. 9. Ottawa: Queen's Printer, 1963. (136 pp.)

Raynauld, A., *The Canadian Economic System*. Translated by C. M. Ross. Toronto: Macmillan, 1967. (440 pp.)

Rea, K. J., *The Political Economy of the Canadian North: An Interpretation of the Course of Development in the Northern Territories of Canada to the Early 1960s.* Toronto: University of Toronto Press, 1968. (453 pp.)

Rimbert, S., "Essai méthodologique sur des stéréotypes régionaux au Canada", *Cahiers de Géographie de Québec*, No. 36, 1971, pp. 523–536.

Ritchot, G., "Problèmes géomorphologiques de la vallée du Saint-Laurent", *Revue de Géographie de Montréal* XVIII (1964): 5–64 and 137–234. See also Vol. XXI (1967): 41–79, 169–189 and 267–311.

Robinson, I. R., *New Industrial Towns on Canada's Resource Frontier*, Dept. of Geography Research Paper, No. 73. Chicago: University of Chicago Press, 1962. (190 pp.)

Robinson, J. L., *Resources of the Canadian Shield.* Toronto: Methuen, 1969. (136 pp.)

Ruggles, R., "The West of Canada in 1763: Imagination and Reality", *The Canadian Geographer* XV (1971): 235–261.

School of Community and Regional Planning, University of British Columbia, *Canada: Isodemographic Map of Canada. Isodemographic Map of Major Cities.* Ottawa: Dept. of Fisheries and Forestry, 1971.

Schwartz, M. A., *Public Opinion and Canadian Identity.* Berkeley: University of California Press, 1967. (263 pp.)

Siegfried, A., *Le Canada, puissance internationale.* Paris: Colin, 1947. (272 pp.)

Stone, L. O., *Migration in Canada. Regional Aspects*, 1961 Census Monograph. Ottawa: Queen's Printer, 1969. (407 pp.)

Tanner, V., *Outlines of the Geography, Life and Customs of Newfoundland-Labrador* (*the Eastern Part of the Labrador Peninsula*), *Acta Geographica*, Vol. 8, No. 1. Helsinki: 1944. (906 pp.)

Taylor, G., *Canada. A Study of Cool, Continental Environments and Their Effects on British and French Settlement.* London: Methuen, 1947. (526 pp.)

Territorial Education System, *Survey of Education. Northwest Territories, 1972.* Yellowknife, 1972. (213 pp.)

Thomson, D. W., *Men and Meridians, The History of Surveying and Mapping in Canada.* 3 vols. (*Prior to 1867*, *1867 to 1917* and *1917 to 1947*). Ottawa: Queen's Printer, 1966, 1967 and 1969. (345 pp., 342 pp. and 370 pp. respectively.)

Tremblay, M. A. and Anderson, W. J., eds., *Rural Canada in Transition.* Ottawa: Queen's Printer, 1966. (415 pp.)

Trudel, M., *Histoire de la Nouvelle-France.* 2 vols. (*1524–1603* and *1604–1627*). Montreal: Fides, 1963 and 1966. (307 pp. and 544 pp. respectively.)

University of Montreal Geography Department, *Revue de Géographie de Montréal* 21 (1967): 5–408.

Urquhart, M. C. and Buckley, K. A. H., eds., *Historical Statistics of Canada*. Toronto: Macmillan, 1965. (672 pp.)

Wade, M., *The French Canadians, 1760–1945*. Toronto: Macmillan, 1955. (1,136 pp.)

Walker, E. R., Johnson, O. and Boswell, W. H., *Meteorological Characteristics of Chinooks in Alberta*. Defence Research Establishment, Suffield Technical Note, No. 227. Ralston, 1968. (30 pp.)

Warkentin, J., ed., *Canada, A Geographical Interpretation*. Toronto: Methuen, 1967. (650 pp.)

Warkentin, J., ed., *The Western Interior of Canada, A Record of Geographical Discovery 1612–1917*. Carleton Library, No. 15. Toronto: McClelland and Stewart, 1964. (308 pp.)

Watson, J. W., *North America. Its Countries and Regions*. London: Longmans, 1964, pp. 240–492. (854 pp.)

Wilson, G. W., Gordon, S. and Judek, S., *Canada: An Appraisal of its Needs and Resources*. Toronto: University of Toronto Press, 1965. (453 pp.)

Wolforth, J. and Leigh, R., *Urban Prospects*. Toronto: McClelland and Stewart, 1971. (196 pp.)

Wonders, W. C., ed., *Canada's Changing North*. Carleton Library, No. 55. Toronto and Montreal: McClelland and Stewart, 1971. (364 pp.)

Wood, W. D. and Thoman, R. S., eds., *Areas of Economic Stress in Canada*. Kingston: Queen's University Press, 1965. (221 pp.)

Canadian Atlases

Blair, C. L. and Simpson, R. I., *The Canadian Landscape; Map and Air Photo Interpretation*. Toronto: Copp Clark, 1967. (172 pp.)

Brouillette, B., Saint-Yves, M. and Reynaud-Dulaurier, G., eds., *Atlas Larousse canadien*. London, 1971. (128 plates and 43 pp.)

Canada, *Atlas of the Northwest Territories, Canada*, with the report of the Carrothers' Commission, Dept. of Northern Affairs and Natural Resources. 2 vols. Ottawa: Queen's Printer, 1966. (155 pp. and 47 pp. respectively.)

———, *Climatological Atlas of Canada*. Edited by M. K. Thomas. Ottawa: Queen's Printer, 1953. (256 pp.)

———, *Topographical Maps of Canada Illustrating Geology and Land Forms*. Ottawa: Queen's Printer, 1964. (42 maps; bilingual.)

———, *Atlas of Climatic Maps*, Series 1 to 10. Ottawa: Queen's Printer, 1967–1970. (30 maps; bilingual.)

———, *County Atlases of Canada. A Descriptive Catalogue*. Compiled by Betty May et al. Ottawa: Queen's Printer, 1970. (191 pp.)

Canada, *The National Atlas of Canada*. Ottawa: Information Canada, 1970–1972. (Bilingual.) See also *Atlas of Canada*, 1906, 1915 and 1957 editions.

Chambers, J. W., Eccles, W. J. and Fullard, H., *Philips Historical Atlas of Canada*. London: G. Philip and Son, 1966. (56 pp.)

Chapman, J. D. and Sherman, J. C., *The United States and Canada, Oxford Regional Economic Atlas*. Toronto and New York: Oxford University Press, 1967. (164 pp.)

Chapman, J. D. and Turner, D. B., *British Columbia Atlas of Resources*. Vancouver: for the British Columbia Natural Resources Conference, 1956. (92 pp.)

Dean, W. G., ed., *Economic Atlas of Ontario*. Toronto: University of Toronto Press, 1969. (113 pp.)

de Varennes, J. and Lavallée, J., *Atlas général Holt*. Montreal and Toronto: Holt, Rinehart and Winston, 1970. (158 pp.)

Gad, G. and Baker, A., *A Cartographic Summary of the Growth and Structure of the Cities of Central Canada*. Toronto Urban Environment Study Research Report, No. 11. Toronto, 1969. (41 pp.)

Gourou, P., Grenier, F. and Hamelin, L.-E., *Atlas du monde contemporain*. 2nd ed. Montreal: ERPI, 1970. (88 plates.)

Grolier of Canada, *Encyclopedia Canadiana*. 10 vols. Montreal: Grolier, 1958.

Kerr, D. G. G., *A Historical Atlas of Canada*. Don Mills: Thomas Nelson and Sons, 1961. (120 pp.)

Marois, C., *Employment Atlas, City and Island of Montreal*. Montreal: Quebec University Press, 1972. (208 pp.; bilingual.)

Quebec, *Atlas of Quebec*. 4 vols. Quebec: for the Ministry of Industry and Commerce, circa 1967. (Bilingual.)

Richards, J. H., ed., *Atlas of Saskatchewan*. Saskatoon: University of Saskatchewan Press, 1969. (236 pp.)

Swithinbank, C., *Ice Atlas of Arctic Canada*. Ottawa: Queen's Printer, 1960. (67 plates.)

Trudel, M., *An Atlas of New France*. Quebec: Laval University Press, 1968. (219 pp.; bilingual.)

University of Alberta Geography Dept., *Atlas of Alberta*. Edmonton: University of Alberta Press, 1969. (158 pp.)

Warkentin, J., ed., *Manitoba Historical Atlas: a Selection of Facsimile Maps, Plans and Sketches from 1612 to 1969*, with annotations by J. Warkentin and R. I. Ruggles. Winnipeg: for the Historical and Scientific Society of Manitoba, 1970. (585 pp.)

Weir, T. R., ed., *Economic Atlas of Manitoba*. Winnipeg: for the Dept. of Industry and Commerce, 1960. (37 plates.)

Weir, T. R., ed., *Atlas of the Prairie Provinces*. Toronto: Oxford University Press, 1971. (31 plates.)

Wilson, C., *The Climate of Quebec. Climatic Atlas.* Part 1. Canadian Meteorological Service Climatological Studies, No. 11. Ottawa: Information Canada, 1971. (44 plates; bilingual.)

Zaborski, B., *Atlas of Landscapes and Settlements of Eastern Canada.* Montreal: Sir George Williams University Press, 1972. (200 pp.; bilingual.)

Periodicals and Serial Publications in Canadian Geography

Albertan Geographer, The, University of Alberta Geography Department. Edmonton: University of Alberta Press, from 1964.

Bulletin, Association des géographes du Québec. Quebec, from 1962.

Bulletin, Canadian Association of Geographers, Education Committee. Montreal, from 1955.

Cahiers de Géographie de Québec, Institute of Geography, Laval University. Quebec: Laval University Press, from 1952.

Canadian Geographer, The, Canadian Association of Geographers. Toronto: University of Toronto Press, from 1951.

Canadian Geographical Journal, Royal Canadian Geographical Society. From 1929.

GÉCET (Groupe d'étude de choronymie et de terminologie géographique), Dept. of Geography, Laval University. Quebec: Laval University Press, from 1966.

Geographical Bulletin. Ottawa: Dept. of Mines and Technical Surveys, Geographical Branch, from 1951 to 1968.

Geographical Papers. Ottawa: Information Canada, from 1950.

Geoscope, Geographers Association, University of Ottawa. Ottawa: University of Ottawa Press, from 1970.

Occasional Papers, Dept. of Geography, University of Western Ontario. London: University of Western Ontario Press, from 1967.

Occasional Papers in Geography, Canadian Association of Geographers, Western Division. Vancouver, from 1959.

Ontario Geography, Dept. of Geography, University of Western Ontario. London: University of Western Ontario Press, from 1967.

Programme of the Annual Meeting of Canadian Association of Geographers. Occasional publications of proceedings, preconference papers, selected essays or abstracts of papers. From 1951.

Publications of the Dept. of Geography, University of Toronto. Research Publications, from 1966; Natural Hazard Research, from 1968; Discussion Papers, from 1969.

Publications in Climatology, Dept. of Geography, McMaster University. Hamilton: McMaster University Press, from 1968.

Revue de Géographie de Montréal, La, Dept. of Geography, University of Montreal. Montreal: University of Montreal Press, from 1947.

Studies in Geography, Dept. of Geography, University of Alberta. Edmonton: University of Alberta Press, from 1971.

TIGUL (Travaux de l'Institut de géographie de l'université Laval). Quebec: Laval University Press, 1952–1969 (first series) and 1970– (second series).

Western Geographical Series, Dept. of Geography, University of Victoria. Victoria: University of Victoria Press, from 1970.

Index